D1622247

MICROWAVE PRINCIPLES AND SYSTEMS

Nigel P. Cook   A.S.E.E, B.S.E.E.

ITT Technical Institute
La Mesa, California

PRENTICE-HALL, Englewood Cliffs, New Jersey 07632

Library of Congress Cataloging-in-Publication Data

COOK, NIGEL P. (date)
   Microwave principles and systems.

   Includes index.
   1. Microwaves. 2. Radar. 3. Artificial satellites
in telecommunication. I. Title.
TK7876.C67 1986   621.381′3   85–16731
ISBN 0–13–581596–7

Printed in the United States of America

10 9 8 7 6 5 4 3 2

Manufacturing buyer: Gordon Osbourne
Cover design: Edsal Enterprises

ISBN 0-13-581596-7 01

Prentice-Hall International (UK) Limited, *London*
Prentice-Hall of Australia Pty. Limited, *Sydney*
Prentice-Hall Canada Inc., *Toronto*
Prentice-Hall Hispanoamericana, S.A., *Mexico*
Prentice-Hall of India Private Limited, *New Delhi*
Prentice-Hall of Japan, Inc., *Tokyo*
Prentice-Hall of Southeast Asia Pte. Ltd., *Singapore*
Editora Prentice-Hall do Brasil, Ltda., *Rio de Janeiro*
Whitehall Books Limited, *Wellington, New Zealand*

This book is dedicated
to my darling daughter Candice,
and beloved wife Dawn
whose continued moral support and assistance
greatly contributed to the timely completion
of this book.

CONTENTS

③ MICROWAVE COMPONENTS   29

④ MICROWAVE ANTENNAS   65

⑤ TWO MICROWAVE APPLICATIONS   93

⑥ *RADAR*

7   *SATELLITE COMMUNICATIONS*   187

ABBREVIATIONS

Abbreviation	Identity
A.F.C.	Automatic Frequency Control
A.M.	Amplitude Modulation
A.P.	Average Power
C	Capacitance
cm	Centimeters
COMSAT	Communications Satellite Corporation
C.R.T.	Cathode Ray Tube
C.T.U.	Central Terminal Unit
dB	Decibel
E	Electric Field
E.B.M.	Electronic Bearing Marker
E.H.F.	Extra High Frequency
F	Frequency
F.D.M.	Frequency Division Multiplexing
F.D.M.A.	Frequency Division Multiple Access
F.M.	Frequency Modulation
F.S.K.	Frequency Shift Keying
GHz	Gigahertz
H	Magnetic Field
H.F.	High Frequency
Hz	Hertz
I.F.	Intermediate Frequency
I.F.F.	Identify Friend or Foe
I.F.O.	Injection Frequency Oscillator
INMARSAT	International Marine Satellite Organization
INTELSAT	International Telecommunications Satellite Organization
I/P	Input

Abbreviation	Identity
IR	Infrared
I.R.U.	Interference Rejection Unit
KW	Kilowatts
L	Inductance
L.F.	Low Frequency
L.N.A.	Low Noise Amplifier
L.O.	Local Oscillator
MARISAT	Marine Satellite Organization
M.F.	Medium Frequency
mW	milliwatts
O/P	Output
P.A.M.	Pulse Amplitude Modulation
P.D.	Pulse Duration
P.D.M.	Pulse Duration Modulation
P.L.	Pulse Length
P.L.M.	Pulse Length Modulation
P.P.	Peak Power
P.P.I.	Plan Position Indicator
P.P.M.	Pulse Position Modulation
P.R.F.	Pulse Repetition Frequency
P.R.R.	Pulse Repetition Rate
P.R.T.	Pulse Repetition Time
P.S.K.	Phase Shift Keying
P.W.	Pulse Width
P.W.M.	Pulse Width Modulation
RADAR	Radio Detection and Ranging
R.F.	Radio Frequency
R.M.	Relative Motion
Rx	Receiver
S.H.F.	Super High Frequency
S.W.G.	Slotted Waveguide
T/B	Timebase
T.C.M.R.	Typical Commercial Marine Radar
T.D.M.	Time Division Multiplexing
T.D.M.A.	Time Division Multiple Access
T.E.	Transverse Electric
T.E.M.	Transverse Electromagnetic
T.M.	Transverse Magnetic
T/R	Transmit/Receive
T.W.T.	Travelling Wave Tube
Tx	Transmitter

Abbreviation	Identity
U.H.F.	Ultra High Frequency
UV	Ultraviolet
+V	Positive Voltage
−V	Negative Voltage
V.H.F.	Very High Frequency
V.L.F.	Very Low Frequency
V.R.M.	Variable Range Marker
XTAL	Crystal

PREFACE

To explain all the intricacies of microwave generation and utilization would require volumes upon volumes of text and would necessitate a reader who has an already extensive knowledge of mathematics and electronics.

In this book I have endeavored to provide an effective blend of microwave theory and practical elements of an operational microwave system. Microwave theory is covered in the first five chapters, in which I have tried to break down the many complex concepts and present them in a straightforward, comprehensive manner. In the latter part of the book, I have chosen radar and satellite communications as two practical applications that utilize microwave frequencies, and described the common concepts and full operation of a typical marine radar and satellite communicator.

Microwaves, as the name implies, are very small waves. The higher-frequency edge of microwave borders on the optical spectrum, which accounts for the fact that microwaves behave more like rays of light than ordinary radio waves. Due to this unique behavior, microwave frequencies are classified and discussed separately from radio waves.

Chapter 1 of this book begins by giving a brief history of microwave frequencies, including the early experiments of Heinrich Hertz and an interesting account of wartime radar development by the British pioneer, Sir Robert Watson Watt. The latter section of this chapter discusses in detail the microwave frequency spectrum, bands, and present-day applications.

Chapter 2 gives an insight into the standard and nonstandard peculiarities of propagation at microwave frequencies, describing how these small waves are polarized and reflected, bounced, scattered, and absorbed on their travels between transmitter and receiver.

"Microwave Components" is the title of Chapter 3, in which some of the unique devices associated specifically with channelling, generating, amplifying, coupling, and detecting microwave frequencies are comprehensively covered.

Microwave antennas are dealt with in Chapter 4, starting with the fundamentals of antenna theory such as gain, beamwidth, matching, resolution, and lobe structure. Following this is a close-up look at the different characteristics, types, and applications of microwave frequency antennas.

Chapter 5 is an introduction before a detailed description of two systems that utilize microwave frequencies. Our first microwave application is a marine radar in which the principles and basic concepts are introduced. Our second application is satellite communications, in which the three main components of a satellite communication system are separately and then collectively discussed.

After the basic introduction to the principles and main blocks of radar in Chapter 5, Chapter 6 begins by breaking up the transmitter, receiver, and display units into a more detailed block diagram. Following this is an introduction to the TCMR (typical commercial marine radar), which has been designed by the author to represent a state-of-the-art marine radar and to extend our understanding to gain a clear picture of what a typical working radar would consist of. After the introduction to the TCMR, the chapter is split into four sections—scanner units, transmitters, receivers, and display units—in which we discuss how the TCMR's scanner, transmitter, receiver, and display unit achieve the desired results, along with other methods and related information concerning each section.

Chapter 7 deals with satellite communications, beginning with the satellite organizations and then the three basic groups, which are INTELSAT, for point-to-point transoceanic communication; INMARSAT, which provides communication between the shore and ships at sea; and U.S. regional satellites for domestic use. The center section of this chapter discusses multiplexing, pulse modulation, shift keying, codes, and the teleprinter before arriving at a complete description of a marine satellite communication system. Frequencies, signal formats, step-by-step procedure, and a block diagram of a typical marine satellite communication system are comprehensively covered in the final section of this chapter.

Following each chapter is a summary and a set of review questions which will help reinforce what has been discussed in that chapter.

A glossary of terms and index at the back of the book may help you to understand and locate certain topics.

A special thanks to the following, whose support and assistance have aided me in this publication endeavor: Frances Ford, Andrew and Anita Gilmour, Patricia Greig, Joseph Mayo, Ronald Retchless, and Hugh Scriven.

Nigel P. Cook

1

THE PAST AND PRESENT

Objectives

After completing this chapter, you will be able to:

1. Describe the history of microwave frequency generation.
2. Explain the development of radar.
3. State where the microwave frequency band begins and ends.
4. List the three bands within the microwave frequency spectrum.
5. List the four basic quantities of an electromagnetic wave.
6. List all the present-day applications of microwave frequencies.

Heinrich Hertz, a young German physicist, was the first pioneer experimenting in the microwave field. He worked with frequencies around 60 MHz, which is below the lowest microwave frequency, but then in his later experiments he extended his boundaries up to 500 MHz before dying at the early age of 37 in 1894, leaving the field of microwave experimentation wide open to other scientists.

In the first decade following Hertz's outstanding accomplishments, the frequency frontier was quickly extended to 75,000 MHz, and twenty years later to 150,000 MHz.

In 1940, Great Britain successfully developed the cavity magnetron, which could be utilized to supply short, high-powered, microwave frequency pulses for radar applications. During that same year the information was given to the Radiation Laboratory at The Massachusetts Institute of Technology, where they proceeded with design of airborne radar equipment.

DEVELOPMENT OF RADAR

Radar was developed in Great Britain at the turn of the century just after Marconi's famous work in wireless communications. Communicating across the English channel, British scientists discovered that at higher frequencies a steel ship reflected the frequency. They also discovered that radio receivers would receive static from electrical storms over the channel. In fact, by adjusting the direction of your antenna one could locate the bearing of the center of the storm, and the intensity of the static would give an indication of the range.

In 1930 Sir Robert Watson Watt, considered to be the "father of ra-

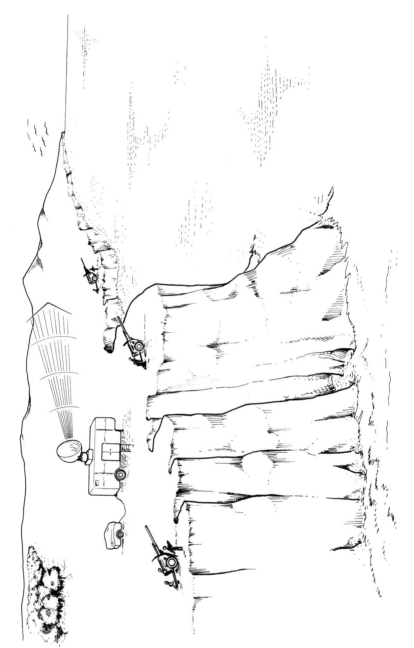

Figure 1-1 Radar van in the field.

dar,'' was commissioned by the British government to design reflected wave detectors. This device was one of the most closely guarded secrets in World War II and was installed on battleships, submarines, and airplanes. Unfortunately, as with any new device there was a reluctance by some commanding officers to use radar. One example of this was the commanding officer of the British battleship H.M.S. *Hood* on his mission to sink the German battleship *Bismarck*. On encountering the *Bismarck,* he chose the distance given by an optical range finder in preference to that of the radar; he failed to hit his target, but the *Bismarck* commander managed to hit the *Hood* with the first salvo.

In 1941, Britain was equipped with many coastal radar stations. Operating at a low frequency of 200 MHz, radar had the ability to see through fog, clouds, and darkness, giving accurate range and tracking of German aircraft. As a result Hitler's bombers during the Battle of Britain took a beating that they never expected.

Large trailers made compact radar equipment mobile, providing the military and civilian airlines with a versatile defense and operational radar system (Fig. 1-1).

By 1943 most of the radar equipment had been replaced with 1500 MHz, microwave frequency, installations that greatly aided the allied success both in the air and at sea.

After World War II, microwave communications using relay stations came into use to convey telegraph, telephone, and television network programs. More and more TV relay stations were installed, linking up all cities to various networks; then satellites were installed, linking up countries.

MICROWAVE FREQUENCIES

The microwave frequency band starts at the highest end of the radio frequency spectrum (VHF) and ends where the optical frequency spectrum begins. See Figs. 1-2 and 1-3.

VLF	LF	MF	HF	VHF	UHF	SHF	EHF	Infrared	Visible	Ultraviolet
Ultrasonic					Microwave			Optical		

Figure 1-2   Frequency spectrum.

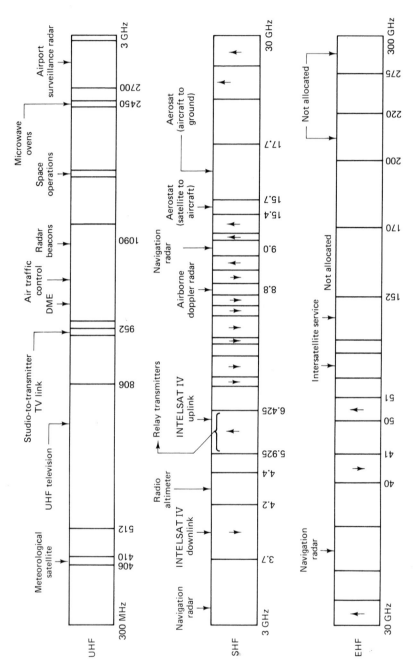

Figure 1-3 The microwave frequency spectrum.

Microwave frequencies include three bands: the ultrahigh frequency (UHF) band of 300 MHz to 3 GHz, the superhigh frequency (SHF) band of 3 to 30 GHz, and the extremely high frequency (EHF) band of 30 to 300 GHz.

When you start to drift into frequencies in gigahertz, (1 GHz = 10^9 Hz = 1,000,000,000 cycles per second), the figures become a trifle cumbersome, so such high frequencies are often referred to according to their wavelength.

ELECTROMAGNETIC WAVES

Microwave energy travels in waves termed *sinusoidal* or *sine waves*. Every wave has the same four basic quantities: (Fig. 1-4)

1. Amplitude
2. Velocity
3. Frequency
4. Wavelength

The amplitude of the sine wave is the magnitude of variation. In our diagram in Fig. 1-4, this is the distance between positive and negative peaks.

The velocity is approximately 300,000,000 meters per second, or 186,000 miles per second for all electromagnetic waves.

Sinusoidal waves can take a long or short time to complete a cycle or alternation. This time period is equal to the reciprocal of the frequency. For

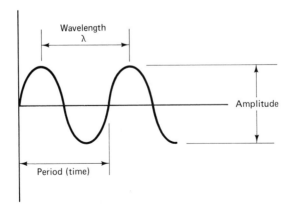

Figure 1-4 Electromagnetic waves.

example, the energy arriving at the home electrical socket alternates at a frequency of 60 cycles per second. This means that 60 waves pass a fixed point in one second or that in one second 60 cycles have been received. The time for one wave to complete its cycle is equal to one-sixtieth of a second or 16.6 milliseconds.

$$T = \frac{1}{f} = \frac{1}{60} = 16.67 \text{ milliseconds}$$

$$T = \text{time in seconds} \qquad f = \text{frequency in hertz}$$

By transposition of the above formula you can calculate the frequency when time is known:

$$f = \frac{1}{T} = \frac{1}{16.67 \text{ ms}} = 60 \text{ hertz}$$

Wavelength, as the name implies, is the physical length of the wave, that is, the distance between two consecutive wave peaks. It is measured in meters per second and is called lambda (λ).

$$\lambda = \frac{V}{f}$$

$$\lambda = \text{wavelength in meters}$$
$$V = \text{velocity of electromagnetic waves in meters per second}$$
$$f = \text{frequency of the wave in hertz (cycles per second)}$$

The microwave band of frequencies span from 0.3 to 300 GHz, which in wavelength is equal to 1 meter to 1 millimeter.

$$\lambda(0.3 \text{GH}_3) = \frac{3 \times 10^8 \text{ m/s}}{0.3 \times 10^9 \text{ Hz}} = 1 \text{ meter}$$

to

$$\lambda(300 \text{GH}_3) = \frac{3 \times 10^8 \text{ m/s}}{300 \times 10^9 \text{ Hz}} = 1 \text{ mm}$$

LETTER DESIGNATIONS FOR MICROWAVE BANDS

Almost all commercial microwave equipment is made to operate in specific bands somewhere between 0.5 to 40 GHz. Refer to Fig. 1-5, which shows the old and new letter designations.

As an example, a previous "X" band marine radar that transmitted a frequency somewhere between 9380 and 9440 MHz would now be named an "I" band radar system.

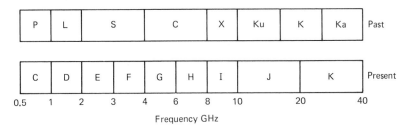

Frequency GHz

Figure 1–5

PRESENT-DAY APPLICATIONS AND FREQUENCIES

Telecommunications (the transmission of analog and digital information from one point to another) is the largest application for microwave frequencies.

Microwave frequencies, being so high, means that a 2 percent bandwidth at a transmitting frequency of 6 GHz is equal to 120 MHz. This large bandwidth makes microwave frequencies ideal for something known as *multiplexing information.* A multiplexed transmission means the simultaneous transmission of two or more signals within a single channel. By the use of multiplex systems, microwave communication systems can be, and generally are, made to carry many telephone conversations at the same time.

The large antenna towers (Fig. 1–6), which can be seen across the country, act as microwave relay stations connecting distant telephone networks, and occupy the 5.925- to 6.425-GHz band.

Radar, which is an acronym for radio detection and ranging, is a very important application of microwave frequencies. A radar system has a transmitter that supplies a very high-power, short-duration microwave frequency pulse to a directional antenna. This pulse travels out, away from the antenna, until it strikes an object. Part of the energy (an echo) is reflected back to a receiver, where it is amplified and then fed to a display unit. The display unit compares the time it took from transmission until the echo was returned. This time allows us to calculate the range of the object, and the direction we received the echo from will tell us the bearing. This is called *primary radar,* when the signals are returned by reflection. Radar equipment operates on 8.5 to 9.2 and 13.25 to 13.40 GHz, and is discussed in greater detail in Chapters 5 and 6.

Aircraft pilots use altimeters that utilize the radar principle for measuring height above the ground. These operate at a frequency of 15.7 to 17.7 GHz.

Operating in the 1030 to 1090 MHz band, IFF (identify friend or foe) is a military radar system used to identify an allied or enemy aircraft or

Figure 1-6   Microwave communication antenna link.

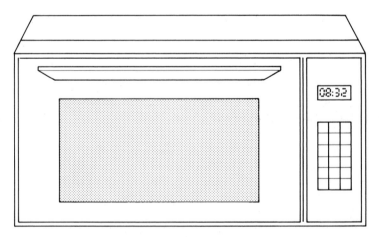

Figure 1-7   Microwave oven.

ships. It is called a *secondary radar system,* because the incident signal from our vessel triggers a responder in the unidentified vessel, whose return coded transmission will enable us to determine friendly from hostile forces.

Satellite communications is evolving into a very large and important application of microwave frequencies. Many commercial and government satellites are in circular orbits approximately 22,300 miles above the earth's equator. The satellites act as microwave relay stations that convey television signals, telephone calls (analog information), and telegraph communications (digital information). Satellite communications are covered in greater detail in Chapters 5 and 7.

One microwave application that can be found in many homes is the microwave oven. Operating at a frequency of 2.45 GHz, it has become very popular amongst consumers for very fast cooking. See Fig. 1-7.

SUMMARY

1. Heinrich Hertz was the first pioneer to experiment in the microwave field.
2. Microwave frequencies exist between 0.3 to 300 GHz, which equals 1 meter to 1 millimeter in wavelength, respectively.
3. Telecommunications is the largest application for microwave frequencies.
4. Radar, which is an acronym for radio detection and ranging, and satellite communications are other large applications that utilize microwave frequencies.

REVIEW QUESTIONS

1. Calculate the wavelength in centimeters for the following frequencies:
   (a) 6 GHz
   (b) 18 GHz
   (c) 127 GHz
   (d) 275 GHz
2. Give the names of three applications that use microwave frequencies.

3. Calculate the time it takes to complete one cycle of an electromagnetic wave at frequencies of:
   (a) 300 MHz
   (b) 300 GHz
   (c) 60 GHz
   (d) 240 GHz

4. Microwave frequencies span three bands. What are these bands named?

5. Wavelengths at microwave frequencies are normally measured in:
   (a) Millimeters
   (b) Centimeters
   (c) Voltmeters
   (d) Microns

6. Describe the difference between primary and secondary radar.

7. Electromagnetic waves travel at a speed of (fill in the blanks):
   (a) _____miles/second
   (b) _____meters/second

8. Every sinusoidal wave has four basic quantities, which are
   (a) _____
   (b) _____
   (c) _____
   (d) _____

9. Give the full names for the following abbreviations:
   (a) EHF
   (b) SHF
   (c) UHF
   (d) IFF
   (e) Hz
   (f) VHF

11. Microwave ovens operate at a frequency of_____.
   (a) 1.37 GHz
   (b) 2.45 GHz
   (c) 3.37 GHz
   (d) 22 MHz

12. IFF is
   (a) An abbreviation for "identify friend or foe"
   (b) A secondary radar system
   (c) An abbreviation for International Federation Frequency
   (d) Both (a) and (b)

13. What are the four basic quantities of an electromagnetic sinusoidal wave?
   (a) _____
   (b) _____
   (c) _____
   (d) _____

14. What is a multiplexed transmission?

15. The microwave band of frequencies spans from_____ to _____.

2

MICROWAVE PROPAGATION

Objectives

After completing this chapter, you will be able to:

1. List the five conditions that determine how microwave energy travels between transmitter and receiver.

2. List the four paths of radio waves.

3. State why the microwave signal bends.

4. Define:

    a. Polarization

    b. Horizontal polarization

    c. Vertical polarization

5. Describe the difference between the geometric, optical, and microwave horizons.

6. List the nonstandard propagation conditions.

It is usually possible to achieve reliable operation at distances of up to 500 miles, with communication equipment operating at frequencies below 30 MHz. However, microwave frequencies, as we will discover in this chapter, act differently and short distance is the rule.

Microwave propagation means the way in which microwave energy travels from one point to another, and this depends on:

1. The height of the transmitting and receiving antenna.
2. The particular frequency being used.
3. The distance between the two points.
4. The terrain.
5. The weather conditions along the path.

Because of weather conditions, which is the variable in the above characteristics, microwave engineers planning a communication link have to predict the worst-case condition and choose a suitable path, equipment, and site that will permit the most reliable operation possible.

THE FOUR PATHS OF RADIO WAVES

Radio waves travelling from one point to another follow four different routes—direct, reflected, sky, and surface waves.

The direct wave, as the name implies, travels directly with no reflections between two points. However, the reflected wave has been bounced off either water or land on its travels between transmitter and receiver. (Refer to Fig. 2–1.)

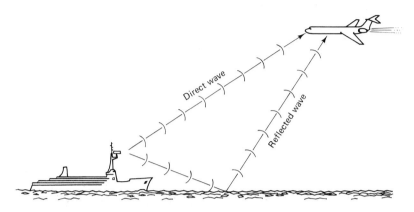

Figure 2-1   Direct and reflected waves.

In most microwave installations we make use of the direct wave only, but in some cases the reflected wave plays a very important role.

Generally, with a reflected wave, a smooth surface, for example, water, gives a clean neat reflection, and a large portion of the energy reaches the target, whereas a land surface reflection gives a multidirection reflection, and therefore only a small amount of energy reaches the target. (Refer to Fig. 2-2.)

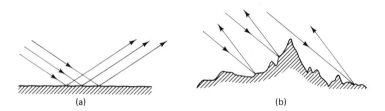

Figure 2-2   Reflections from smooth and rough surfaces. (a) Smooth surface; (b) rough surface.

The field strength at the receiving antenna is the result of adding the direct and reflected wave energy. The reflected wave, therefore, plays an important role when communicating between two points.

If the reflected wave has a longer distance to travel before reaching the receiving antenna, a phase difference between the direct and reflected wave occurs. When the path difference is an odd number of half-wave-lengths, the two signals arrive at the target or receiving antenna in phase; when the path difference is an even number of half-wavelengths, the two arrive out of phase and cancel one another, as illustrated in Figure 2-3.

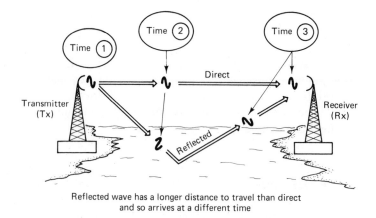

Reflected wave has a longer distance to travel than direct
and so arrives at a different time

When reflected wave returns after an odd number of half wavelengths of the direct wave:

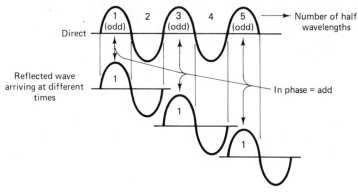

When reflected wave returns after an even number of half wavelengths of the direct wave:

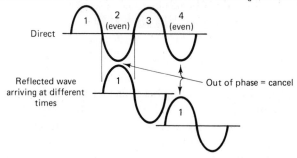

Figure 2-3   In phase and out of phase reflection.

The sky wave has no importance in high-frequency microwave propagation, as reflection from the ionosphere diminishes as frequency increases.

Surface waves travel along the surface of the earth or water. The
transmitting antenna must be very low, within a few wavelengths of the
ground, to produce a surface wave and, as we will discover later, for microwave frequencies to achieve long distances, the transmitting antenna
must be as high as possible.

BENDING OF THE MICROWAVE PATH

The speed at which a microwave signal passes through the atmosphere depends on temperature, precipitation (moisture in the air), and atmospheric
pressure. Generally speaking, the signal travels faster when temperature is
higher, atmospheric pressure is lower, and water vapor is less. The result
you can gather from the previous statement is that the signal speed changes
with altitude. To be specific, as you go up in altitude, the density of the air
decreases resulting in the top of the wave travelling faster, causing a downward bending of the wave. (Refer to Fig. 2-4.)

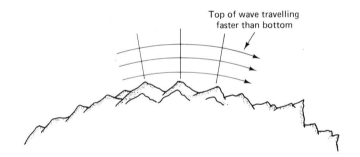

Figure 2-4   Curvature of microwave signal path.

POLARIZATION

An electromagnetic wave propagating between transmitter and receiver consists of an electric or voltage field and a magnetic field, which are always
perpendicular to one another. By definition, the polarization of an electromagnetic wave is the direction of the electric field. If the electric field is
perpendicular to the earth, the wave is said to be vertically polarized; if the
electric field is horizontal to the earth, the wave is horizontally polarized.

An antenna, as we discover in Chapter 4, can be thought of as a matching device between a transmission line and free space, which also directs the radiated field. The antenna actually determines the polarization of the radiated field. A horizontally polarized antenna launches a wave that is absent of any vertical components, and therefore such an antenna will not receive any vertical components. Consequently, for maximum efficiency the receiving antenna must have the same polarization as the transmitting antenna.

As just mentioned, a microwave or lower-frequency radio wave consists of two electrical fields (electric and magnetic), each of which is related to the other in the following ways:

1. They exist in the same region of space.
2. Both move together at the same speed in the direction of propagation.
3. At any point along the direction of propagation, each field has a direction and signal strength.
4. At every instant, the electric and magnetic fields are at an angle of 90 degrees to each other, and both are 90 degrees to the direction of propagation.

The vectors in Fig. 2–5 indicate both direction and strength of the field. Similarly, each broken arrow represents the magnetic field and each solid arrow the electric field.

The entire field of the microwave signal contains an infinite number of wavefronts, all moving in the same direction and at the same velocity. If we could take a knife and cut at any point along the direction of propagation, we would have a wavefront with the electric (E) field and magnetic

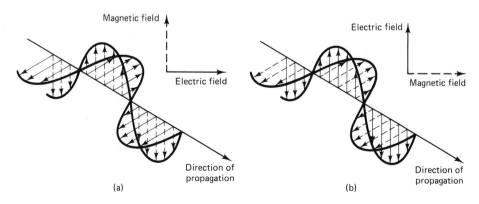

Figure 2-5   (a) Horizontally polarized wave; (b) vertically polarized wave.

(*H*) field both at the same amplitude at the same time (in phase), but perpendicular to one another.

Figures 2–6(a) through 2–6(c) show an electromagnetic wave being radiated from the transmitting antenna towards the receiving antenna, which is one mile away. Let us consider the wavefront at point (A). All electromagnetic waves travel at approximately the same velocity of 186,000 miles per second, so our wavefront will be travelling for one mile at a speed of 186,000 miles per second. The time it will take will be equal to:

$$\text{Time} = \frac{\text{range in miles}}{\text{velocity}} = \frac{1 \text{ mile}}{186,000} = 5.3 \ \mu\text{s}$$

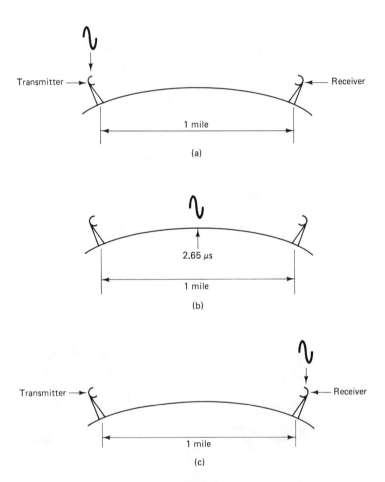

(a)

(b)

(c)

Figure 2–6

The understanding you should gain from this discussion on polarization is that electromagnetic waves are made up of an electric and a magnetic field which both travel in the direction of propagation. A horizontally polarized antenna radiates a horizontally polarized wave that can be received only by a horizontally polarized receiving antenna. A wavefront is a moving point travelling in the direction of propagation.

MICROWAVE HORIZON

Figure 2-7 illustrates the geometric (straight line) horizon A, the slightly curved optical horizon B, and the microwave horizon C. Under standard propagation conditions the microwave horizon is found to be 6 percent more than the optical horizon; this is due to microwaves being a lower-frequency radiation than light waves and therefore bending more.

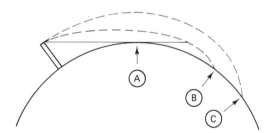

Figure 2-7 Horizons.

Let us now take an example of a marine radar system and calculate the radar horizon. To assist in calculations and the drawing of diagrams, the earth is assumed to be four-thirds its actual radius. Using this value, we can draw the slightly curved microwave path as a straight line; calculations based on straight-line paths make it easier to acquire a formula for calculating the radar horizon.

Figure 2-8 illustrates this idea. The height of the transmitting radar antenna is placed into the formula to calculate the radar's horizon, and therefore its maximum range if sufficient transmitting power is available. If the target's height is high enough to be hit by the transmitting wave, targets can be detected at a distance of A (radar horizon) + B (target's distance beyond radar horizon).

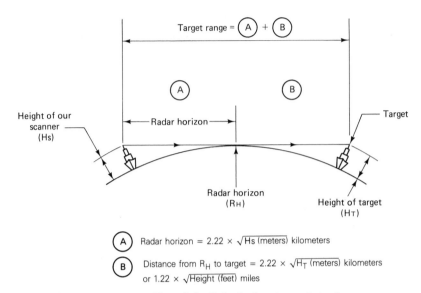

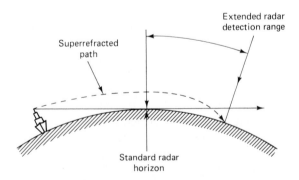

Figure 2-8   Radar horizon in nautical miles.

NONSTANDARD PROPAGATION

1. Super-refraction—frequently occurs in the tropics where very cold sea currents are predominant, for example, the gulf of Aden, or where warm land winds come out over a cold sea, for example, the Red Sea, the Mediterranean Sea, and the English Channel. This results in an increased detection range due to greater bending effect on the wave. See Fig. 2-9.

2. Ducting—is a severe case of super-refraction, in which warm dry air from a land mass flows over a cold sea, and a temperature inversion becomes established. As soon as this dry warm air moves over the sea, a large evaporation into the lower layers of the air occurs, and ducts are formed. Duct-

Figure 2-9   Super-refraction.

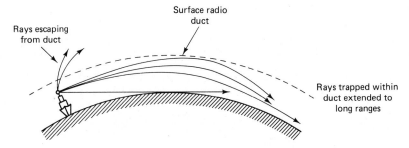

Figure 2-10   Ducting.

ing or severe super-refraction causes an even greater extended detection range. See Fig. 2-10.

3. Skip Effect—Figure 2-11 illustrates a skipping action due to the gradual sinking down and spreading out of air at high altitudes.

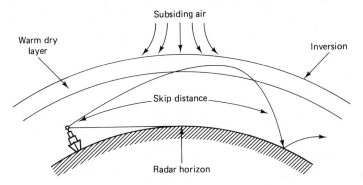

Figure 2-11   Skip effect—greatest extended detection range.

4. Subrefraction—is a detrimental effect on our microwave transmission and occurs in polar regions, where arctic winds blow over warm sea currents. If this condition is prevalent, which is the opposite to the preceding three, it causes a reduced range of as much as 40 percent. See Fig. 2-12.

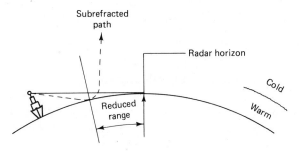

Figure 2-12   Subrefraction.

EFFECTS OF PRECIPITATION (moisture from the sky)

Precipitation causes both absorption and scattering of the microwave energy, and this attenuation increases as frequency increases. Small drops of water, fine particles of dust and sand, hail, snow, or ice in the air act as reflectors that scatter the energy in all directions. This effect becomes increasingly worse as frequency increases, because the particles are now a greater size with respect to the wavelength. See Fig. 2-13.

Fog, sand and dust storms	Hail and snow	Rain
Slight attenuation, does not usually cause a reflection	Moderate attenuation, causes scattering and reflections	Heavy attenuation and strong reflections

Figure 2-13   Precipitation.

At certain frequencies, the atmosphere will not transmit microwaves, but will instead absorb most of the energy. The effect has nothing to do with reflection; instead, the energy is absorbed in precisely the same way that red light is absorbed by a pair of green sunglasses.

Because radar signals are reflected from raindrops, storm centers can be detected easily. The National Oceanic and Atmospheric Administration extensively uses radar systems for detecting and ranging storm centers.

PROPAGATION EFFECTS ON COMMUNICATIONS INSTALLATIONS

We have already mentioned that the reflected signal may be neglected in an overland communication installation, but must be seriously considered in a overwater installation. See Fig. 2-14.

If the direct path nearly touches the surface of the sea, the direct and reflected signal paths are the same length. The reflected signal suffers a phase reversal on reflection, so it will arrive at the receiving antenna precisely out of phase with the direct signal, and therefore an almost complete cancellation will result. To solve this problem either or both antennas are made higher so that the length of the reflected path is increased. If the increase in height is chosen carefully, the length of the reflected path will be one-half wavelength longer than the direct path, causing a phase differ-

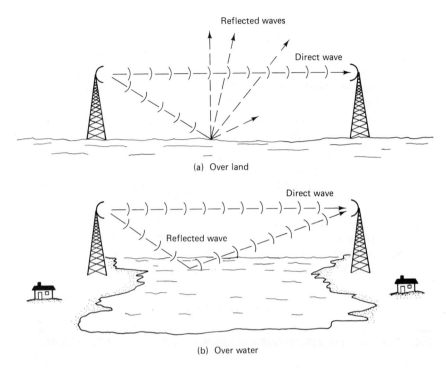

Figure 2-14 A microwave communication link joining two cities. (a) Over land; (b) over water.

ence of 180 degrees between the direct and reflected signals. Therefore, this 180-degree phase difference is added to the 180-degree phase difference brought about by phase reversal on reflection, thus causing the two signals to arrive in phase and add to each other.

When one is designing such a microwave communication link, it is necessary only to determine whether an unobstructed microwave straight line of sight connects the two points between which the microwave operation is desired. This can best be achieved by visual inspection.

SUMMARY

1. At microwave frequencies, communication between two points is largely dependent on the direct wave, so communication or radar detection is impossible if the microwave energy is in any way obstructed.

2. Microwave frequencies suffer a downward bending due to the top of the wavefront's travelling faster than the bottom. This downward bending means that our wavefront hugs the curvature of the earth on its travels between *A* and *B*. The greater the frequency, the less the bending, and so our optical horizon falls a lot short of our microwave horizon.

3. In some conditions the microwave signals can be greatly extended (super-refraction, ducting, and skip effect), or the curvature can be reduced by an upward bending (subrefraction) which greatly reduces the microwave horizon.

4. Precipitation causes absorption and scattering of the microwave energy.

5. Propagating electromagnetic waves are either horizontally or vertically polarized. The plane of the electric field determines whether the wave is horizontally or vertically polarized; that is, if the electric field is vertical to the earth, the wave is said to be vertically polarized.

REVIEW QUESTIONS

1. Radio waves follow four different paths when travelling between two points; these are
   (a) _____
   (b) _____
   (c) _____
   (d) _____

2. Which of the two in Question 1 are not normally present at microwave frequencies?

3. The field strength at the receiving antenna is the result of adding which of the two waves from Question 1?

4. If the reflected wave has to travel one-half wavelength more than the direct wave, the two waves will arrive at the receiving antenna:
   (a) In phase
   (b) Out of phase
   (c) At exactly the same time
   (d) The waves will not arrive

5. Under standard propagation conditions, the top of the wavefront tends to travel _____, causing a _____ bending of the wave.
   (a) Slower, upwards
   (b) Faster, upwards
   (c) Faster, downwards
   (d) Slower, downwards

6. A communication installation across water has been designed, but the microwave engineers have discovered that the direct and reflected waves arrive out of phase, causing cancellation of the signal. To resolve this problem they could:
   (a) Introduce a sky wave
   (b) Remove the ground wave
   (c) Increase the height of either, or both antennas
   (d) Increase the distance between the antennas

7. The success or failure of a microwave communication link will depend on the presence or absence of:
   (a) The direct wave
   (b) The reflected wave
   (c) The sky wave
   (d) The ground wave

8. Generally speaking, the higher the temperature, the lower the atmospheric pressure; the less the amount of water vapor, the:
   (a) Faster the signal travels
   (b) Slower the signal travels

9. The atmospheric effect which reduces signal strength due to the presence of small drops of water or dust particles is known as:
   (a) Ionospheric losses
   (b) Lagging
   (c) Scattering
   (d) Impedence

10. Microwave repeater stations are primarily needed:
    (a) Because microwaves alternate rapidly
    (b) To increase bandwidth
    (c) Because microwaves do not follow the curvature of the earth
    (d) They are not needed

11. An electromagnetic wave propagating between two points consists of an _____ and a _____ field.

12. A horizontally polarized antenna will launch a _____ polarized wave.

13. The electric and magnetic fields of an electromagnetic wave are related to one another in that:
    (a) They move at the same speed in the same direction.
    (b) They are always perpendicular to one another.
    (c) They are both perpendicular to the direction of propagation.
    (d) All of the above
    (e) None of the above

14. The geometric horizon is:
    (a) Longer than the microwave horizon.
    (b) Longer than the optical horizon.
    (c) Shorter than both horizons.
    (d) None of the above

15. Which of the following nonstandard propagation conditions causes a reduction in range?
    (a) Super-refraction
    (b) Skip effect
    (c) Subrefraction
    (d) Ducting

16. Precipitation causes:
    (a) Absorption of microwave energy
    (b) Does not affect microwave frequencies
    (c) Scattering of microwave energy
    (d) Both (a) and (b)
    (e) Both (a) and (c)

17. Calculate the radar horizon (in meters) if the scanner is placed at a height of 20 meters.

18. List the three nonstandard propagation conditions that cause an increase in radar range:
    (a) _____
    (b) _____
    (c) _____

19. The speed at which a signal passes through the atmosphere is dependent on:
    (a) _____
    (b) _____
    (c) _____

20. The electric and magnetic fields of an electromagnetic wave are related to one another in four ways, which are
    (a) _____
    (b) _____
    (c) _____
    (d) _____

3

MICROWAVE COMPONENTS

Objectives

After completing this chapter, you will be able to:

1. State why waveguides are used at microwave frequencies.
2. Explain the development of the rectangular waveguide.
3. Define:
   a. Cutoff
   b. Dominant mode
   c. Transverse electric mode
   d. Transverse magnetic mode
4. State the two input and output probe coupling methods.
5. Describe the operation and function of waveguide dummy loads and variable attenuators.
6. Explain the development of cavity resonators.
7. State the difference between a microwave frequency bandpass and bandstop circuit.
8. Identify the most common devices used for amplification and oscillation at microwave frequencies.
9. Describe how microwave frequencies are mixed and detected.

WHY WAVEGUIDES?

When discussing radio communication systems at lower frequencies, we think of radio frequency energy travelling along transmission lines as voltage and current, and then being transmitted into free space via an antenna, where the energy is now referred to as electromagnetic waves.

Transmission lines such as open-wire lines, coaxial cable, and twin lead lines are generally not used at high microwave frequencies because of large losses brought about by *skin effect*. Skin effect, also called *radio frequency interference,* is the tendency of RF currents to flow near the outer edge of the conductor (the skin). Thus they are restricted to a small part of the total sectional area, which has the effect of increasing the conductor's resistance as you go up in frequency. It can be assumed, therefore, at very high microwave frequencies, that any energy injected onto a transmission line would not even travel along the skin of the conductor, but would in fact just radiate off into free space, so some other conductor of microwave energy had to be found.

Because microwave frequencies will not travel along conductors as current but prefer to travel as electromagnetic waves, we had to find a device that could channel this electromagnetic energy in the direction we desire. A waveguide, is a hollow metal conductor that can be rectangular or circular in cross section, and is used to guide our electromagnetic waves. See Fig. 3-1.

Waveguides are normally made of brass or aluminum to be rustproof and come in sections of various lengths. They may be bent to some angle, twisted to some desired angle, or even made completely flexible. At each end of the waveguide section is a metal flange to allow one section to be bolted to the other; see Fig. 3-1.

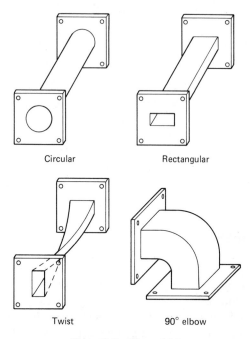

Circular   Rectangular

Twist   90° elbow

Figure 3-1   Waveguides.

For the sake of discussion as to why waveguides are needed, let us consider again exactly where the microwave band of frequencies is.

It lies between the end of the radio frequency band and the beginning of the optical frequency spectrum. In fact, as we discussed in Chapter 2, microwave frequencies are unique in their behavior; as we become better acquainted with microwaves, we will discover that these small frequencies are radio waves that tend to act like light waves.

Let's take the last statement and apply it to an analogy. If we took a flashlight and shone it onto a two-wire transmission line, you can easily imagine that none of the light energy would travel down the line. But if we took that same flashlight and fed it into a section of waveguide, the electromagnetic light energy would emerge out of the end of the waveguide section. This simple analogy may help you to conceptualize why waveguides can be used to guide high-frequency microwave energy.

DEVELOPMENT OF THE RECTANGULAR WAVEGUIDE

Although a waveguide may have any shape, the most common type in use today is rectangular in shape.

The dimensions of the waveguide are always the inside measurements,

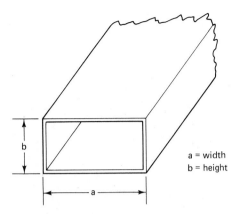

Figure 3-2 Rectangular waveguide dimensions.

with the smaller dimension being taken as the height, even though it may line up in the horizontal plane. See Fig. 3-2.

Figure 3-3 illustrates a two-wire transmission line for a piece of communications equipment at radio frequencies.

Insulators A and B may be regarded as some value of impedance to ground, and this impedance would obviously have to be very large in order to ensure that none of the energy travelling down the conductor is shunted away from the transmission line, causing a reduction in signal strength.

At the lower radio frequencies, the line can be supported by using the insulators as shown in Fig. 3-3, or one could utilize another practical way to support the transmission line.

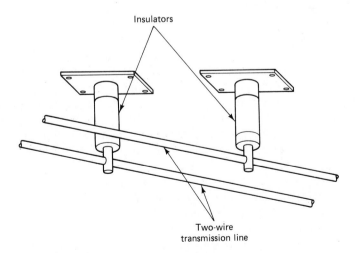

Figure 3-3 Two-wire transmission line with insulators.

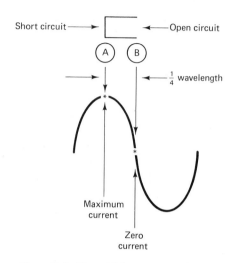

Figure 3-4   Sinusoidal current waveform.

Figure 3–4 shows a sinusoidal current waveform similar to that travelling down our transmission line. If current is maximum at point A (maximum current occurs at a short circuit), then a quarter of a wavelength away
from that point, point B, current will be zero (zero current occurs at an
open circuit).

If you take a short-circuited, quarter-wave section, as seen in Fig. 3–
4, there would be maximum current where the quarter-wave section joins
the transmission line, and a quarter of a wavelength away from this maximum current would be an open circuit or high impedance point. This section could therefore act as a metallic insulator and could be used to support
transmission lines; the two-wire transmission line would be suspended by a
quarter-wave metal rod from a metal wall. See Fig. 3–5. A metal wall or
shorting bar, a quarter-wavelength away from the transmission line, acts
as a short circuit; therefore an open circuit would be presented a quarter-
wavelength away at the junction where the transmission line is joined to
the quarter-wavelength support rod. No energy would be shunted away from
the transmission line by the metal rod, because that path displays a high
impedance, so no loss of power results.

Quarter-wave metallic insulators are more practicable at microwave
frequencies because wavelengths are normally measured in centimeters, so
quarter-wavelength supports are very small. However, at radio frequencies,
the wavelength becomes much larger and quarter-wavelength rods could be
used, but their large size makes them cumbersome to work with.

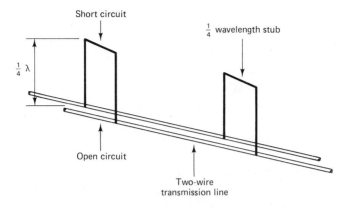

Figure 3–5 Two-wire transmission line supported by quarter-wavelength stub.

In Fig. 3–6 we have carried our example one step further, but now our two-wire transmission line is being supported both at the top and bottom by two quarter-wave sections connected together to make half-wave frames.

If we increase the number of frames, we increase the conduction properties of the two-wire transmission line, and our conducting device now begins to resemble a section of rectangular waveguide. Of course, a wave-

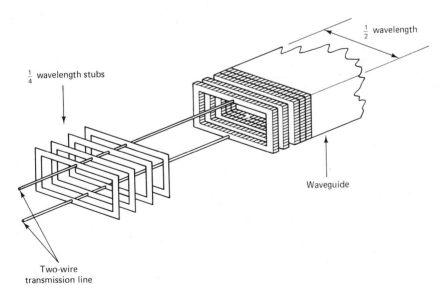

Figure 3–6 Waveguide development.

guide is not constructed by welding together half-wave frames, but instead is built as a single piece.

CUTOFF FREQUENCY OF A WAVEGUIDE

A two-wire transmission line will conduct anything from a high VHF frequency all the way down to 0 Hz, direct current.

Waveguides, however, act as high-pass filters having a low-frequency cutoff that is determined by the physical dimensions of the waveguide. As you already know, a frequency decrease causes wavelength to increase; that is, wavelength is inversely proportional to frequency:

$$\uparrow \lambda = \frac{V}{f} \downarrow$$

When the frequency becomes too low, the metal case of the waveguide will no longer be equivalent to quarter-wavelength stubs on either side of the transmission line, and the energy propagating down the waveguide will be shunted away from the main signal path, as the shunt path no longer presents a high impedance. The waveguide consequently acts as a high-pass filter, passing only cutoff frequencies and above.

Table 3–1 lists some waveguide sizes and their frequency ranges.

TABLE 3–1

Frequency range, GHz	Waveguide size, in.
1.12–2.7	6.5 × 3.25
2.6–3.95	3.0 × 1.50
3.95–5.85	1.87 × 0.87
4.9–7.05	1.59 × 0.795
5.85–8.2	1.37 × 0.62
7.05–10.0	1.12 × 0.497
8.2–12.4	0.9 × 0.4
10.0–15.0	0.75 × 0.375
12.4–18.0	0.62 × 0.31
15.0–22.0	0.51 × 0.255
18.0–26.5	0.42 × 0.17
26.5–40.0	0.28 × 0.14

The lowest frequency listed is the cutoff frequency that will be propagated down the waveguide. However, all frequencies below cutoff, for each particular size of waveguide, will be greatly attenuated.

MODES OF TRANSMISSION IN A WAVEGUIDE

The mode describes the various electromagnetic energy patterns that are propagated down the interior of a waveguide.

In the transverse electric mode (TE), the electric field is transverse (crossing) the direction of propagation, and the magnetic field is parallel to the direction of propagation. Figure 3-7 illustrates a transverse electric

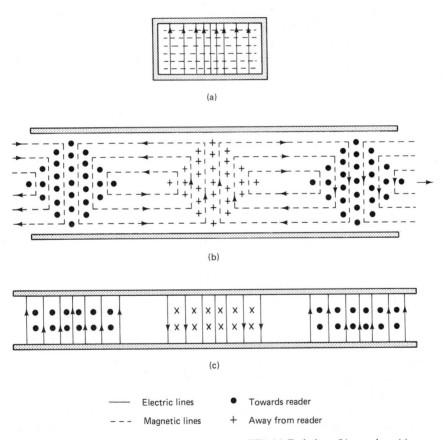

(a)

(b)

(c)

——— Electric lines   ● Towards reader

– – – Magnetic lines   + Away from reader

Figure 3-7   Transverse electric mode pattern (TE). (a) End view; (b) top view; (c) side view.

mode pattern propagating down a rectangular waveguide. The solid lines indicate the electric lines of force and the dotted lines, the magnetic. The black dots (●) and plus signs (+) show the direction of the electric field, be it travelling into or out of the page in the top view. The dot and the plus sign are actually the tip and end of an imaginary arrow which is used to illustrate direction. The dot or head of the arrow indicates that line of force is coming toward you, and the tail demonstrates that the line of force is travelling away from you. The side view uses the same system, but now the dot and + are used to illustrate magnetic lines of force in the side view of a rectangular waveguide. Figure 3–8 combines the side and top view so that one can gain a concept of the two fields perpendicular to one another in the rectangular waveguide. The direction of propagation is as seen by the large arrow travelling down the waveguide from left to right, and you will notice that it does not cut or cross any magnetic lines of force, but in fact fits between two patterns and travels in parallel with them. However, the electric field as seen in the diagram is actually crossing the direction of propagation arrow, so consequently this mode is referred to as *crossing* or *transverse electric* (TE).

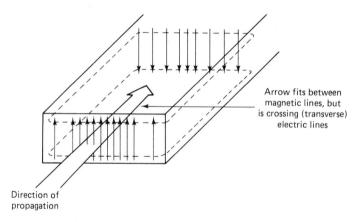

Arrow fits between magnetic lines, but is crossing (transverse) electric lines

Direction of propagation

Figure 3–8   TE mode.

Figure 3–9 shows the side and end view of a transverse magnetic mode propagating down a rectangular waveguide. As can be seen in the side view, the electric lines of force leave the top of the waveguide, proceed down the waveguide, and then return back to the top; conversely, they leave the bottom of the waveguide, proceed along and then return to the bottom. Our direction of propagation arrow travels between two electric lines of force and travels in parallel with these lines. Conversely, our magnetic lines of force are crossing our direction of propagation arrow, so this mode is referred to as *crossing* or *transverse magnetic* (TM).

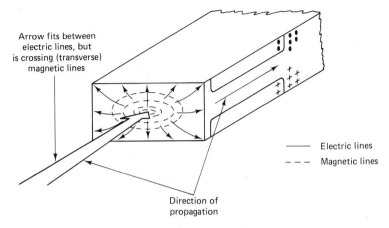

Arrow fits between
electric lines, but
is crossing (transverse)
magnetic lines

———— Electric lines

– – – Magnetic lines

Direction of
propagation

Figure 3–9   TM mode.

DOMINANT MODE OF A WAVEGUIDE

The dominant or best mode of operation in a waveguide is at the lowest frequency when the width of a rectangular waveguide is equal to half a wavelength.

The TE_{10} is the dominant or best mode for propagating energy with minimum losses in a rectangular waveguide. The two subnumerals following TE or TM indicate the number of half-wavelength patterns that fall along the width and height of the waveguide. Our example of TE_{10} states that there is one half-wavelength pattern in the *a* dimension or width of the guide, and uniform or zero half-wavelength patterns in the *b* dimension or height of our rectangular guide.

When a TE mode is discussed, the two subscripts refer to the number of electric field patterns; for the TM mode, we are referring to the amount of magnetic field patterns.

WAVEGUIDE JOINTS AND BENDS

Bends and joints can cause various mechanical complications in a microwave system. Flange joints are therefore smooth and snug-fitting, and bends are machined to have a large radius of curvature to ensure no reflections of our microwave signal. It is necessary to carefully plan and provide a smooth transition between one section of waveguide and the next along the path of propagation. See Fig. 3–10.

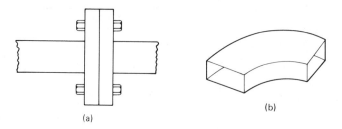

Figure 3-10   Waveguide (a) joint and (b) bend.

INPUT AND OUTPUT PROBE COUPLING

Sometimes it is not desirable to couple microwave energy for a very short distance by using waveguides connected as an elaborate plumbing system. For short distances coaxial cable can be utilized to input or output information to or from a waveguide.

Figure 3-11 illustrates that the inner conductor of a coaxial cable extends like a probe into the waveguide. Its electric field is parallel to the probe, therefore developing a transverse electric mode pattern. As usual, the current flowing up the inner conductor sets up a magnetic field around the conductor that is always perpendicular to the electric field. In this arrangement the waveguide has been excited into the TE_{10} mode.

The small probe is actually inserted a quarter of a wavelength away from the waveguide end wall. When the probe is energized, it actually radiates signals to the left and to the right. The quarter-wavelength distance ensures that the wave radiated to the left has to travel a quarter-wavelength

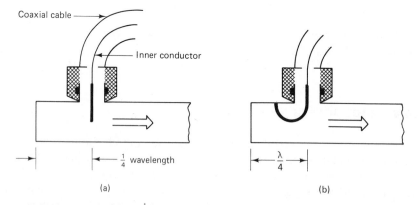

Figure 3-11   Coupling coaxial line to waveguide. (a) Antenna probe; (b) loop or hook.

to the wall and a quarter-wavelength back, a total distance of half a wavelength, or 180 degrees. In fact, the signal radiated to the left actually returns one whole wavelength later (360 degrees) due to the 180 degrees of travel and 180 degrees caused by reflection off the wall. Therefore, the signal radiated to the left will return to reinforce rather than cancel the signal radiated to the right. See Fig. 3–12.

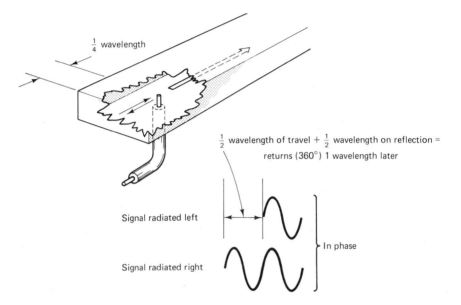

Figure 3–12   In-phase coupling.

A probe extended into a waveguide, as illustrated in Figure 3–13 can actually be used to output a signal from the waveguide. The electric field will cut and induce a voltage into the probe, causing a signal current to flow down the probe out of the waveguide.

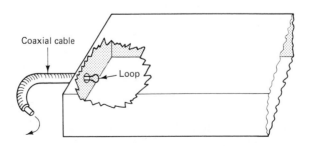

Figure 3–13   Output coupling using loop.

Slit coupling is achieved by small slits actually cut into the walls of the cavity where the *E* field is maximum, that is, the center of the *a* dimension. See Fig. 3–14.

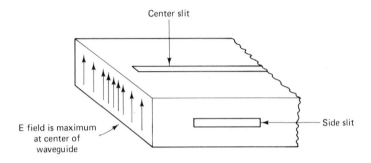

Figure 3-14 Slot or aperture coupling.

WAVEGUIDE ATTENUATORS

For test purposes it may be desirable to dissipate some or all of the energy within a waveguide.

A dummy or artifical load displays the characteristics of an antenna but is nonradiating; that is, it consumes all energy fed to it. Figure 3–15 illustrates the tapered pyramid load, whose pyramid shape ensures no reflections back into the waveguide.

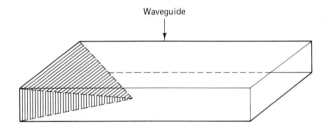

Figure 3-15 Pyramid-type resistive load.

Figure 3–16 illustrates the resistive vane-type dummy load, which achieves the same results of dissipating all energy and allowing no reflections. Both the pyramid and vane dummy loads are coated with a resistive material which, when high power is applied down the waveguide, develop a large amount of heat. Dummy loads designed to consume large amounts of power are normally oil- or water-cooled.

When microwave equipment is tested, a problem arises when the test

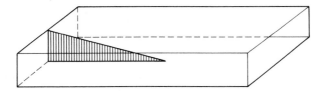

Figure 3-16   Vane-type resistive load.

equipment operates on inputs in milliwatts and the equipment being tested radiates an average power in watts. The test equipment would undoubtedly burn out, or it would disconnect itself and no longer monitor the input power. To eliminate the problem, variable attenuators that use a dielectric coated with resistive material are utilized, such as those seen in Fig. 3-17.

In both diagrams the dielectric is placed in the path of the E field, causing attenuation and therefore power losses. The total loss is dependent on the attenuator's length, the amount of insertion into the waveguide, and

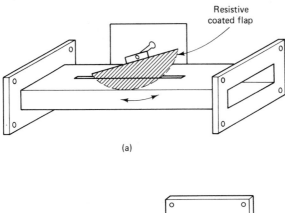

Resistive coated flap

(a)

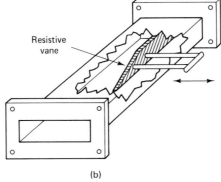

Resistive vane

(b)

Figure 3-17   Variable attenuators. (a) Flap type; (b) vane type.

the resistive material used. These attenuators decrease the signal from no decrease (zero decibels), to more than one-thousandth of the power (30 decibels).

CAVITY RESONATORS

Conventional low-frequency resonant circuits consisting of the usual coil and capacitor are not practical at microwave frequencies, because the required values of capacitance and inductance are extremely small. For this reason and others, which will be explained, it becomes necessary to select a different form of resonant circuit.

When scientists first tried to develop a circuit or component to resonate at higher frequencies, they realized that a hollow, closed metal box of nearly any shape or size could be made to resonate electrically in the same manner as a conventional LC circuit at lower frequencies.

As is generally known, the higher the frequency, the smaller the coil dimensions. In fact, in order to achieve an inductance value for microwave frequencies, the number of turns in the coil is reduced to one, and the capacitance is only that between the two ends of the coil wire. This forms what is know as a hairpin tank; it is illustrated in Fig. 3–18.

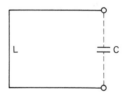

Figure 3-18   Inductance (L) and capacitance (C) of a hairpin tank.

If another hairpin is placed in parallel with our original, as can be seen in Fig. 3–19, in an attempt to reduce the amount of inductance by now having two parallel inductors, the effect is to reduce the inductance but

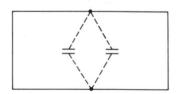

Figure 3-19   Two paralleled hairpins.

unfortunately to increase the capacitance value, therefore leaving this circuit resonant to the same frequency. This is not a problem, because we have already achieved the high frequency we desire with the single hairpin tank first selected, but in fact this paralleling does have the advantage of increasing the Q or quality of the resonant circuit.

If many hairpins are connected to a central point, a cavity similar to a small circular can results, as seen in Fig. 3-20. The size or dimensions of the cavity determine how much inductive and capacitive affect are present, and consequently determine the cavity's frequency.

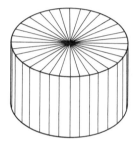

Figure 3-20   Infinite hairpins forming a cavity.

A larger cavity would have a greater amount of capacitance and inductance, and as frequency of resonance is inversely proportional to the LC value, the large cavity would resonate at a lower frequency; a small cavity would resonate at a higher frequency.

By installing a variable plunger, as seen in Fig. 3-21, we effectively vary the distance between the two ends of our hairpin tanks, and therefore the amount of capacitance and, consequently, frequency.

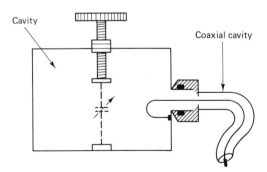

Figure 3-21   Capacitance tuning of a cavity.

Figure 3–22 shows a bandpass cavity that will reject all frequencies above and below its frequency of resonance. The circuit is connected as a series-resonant cavity, as energy has to go through the cavity via the small holes, or apertures. At resonance the cavity, just like any other tuned circuit, will allow maximum energy to pass through. However, frequencies other than resonance will be consumed by the cavity with little or no energy emerging.

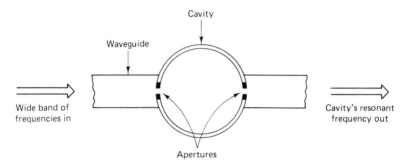

Figure 3-22   Series resonant circuit.

Figure 3–23 shows a bandstop cavity that will pass frequencies on either side of resonance, but will absorb and dissipate energy at the frequency of resonance. The circuit acts as a parallel-resonance circuit, and energy at the cavity's frequency of resonance will be shunted through the aperture into the cavity and away from the main propagation path. A tuning screw is used to move a disk up or down to change the dimensions of the cavity to the desired frequency to be consumed.

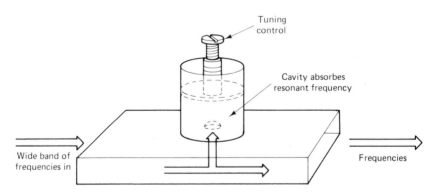

Figure 3-23   Parallel resonant circuit.

MICROWAVE VACUUM TUBES

During the course of our discussions, we will highlight three vacuum tubes that are used exclusively in microwave systems as microwave oscillators and amplifiers.

A magnetron oscillator is used to supply short-duration pulses at megawatts of peak power for radar transmitters.

A klystron oscillator or amplifier is used to amplify a microwave signal or to supply low-power, continuous wave output that is generally used as a local oscillator in a microwave superheterodyne receiver.

The travelling-wave tube (TWT) is generally used as a linear amplifier of microwave signals in earth and satellite communications equipment.

Magnetrons. Refer to Fig. 3-24, which shows the construction of a magnetron oscillator. The magnetron is a diode with an anode and a heated cathode. The anode is actually a metal block with small machined cavities whose size, as we discussed before, determine the frequency of oscillation.

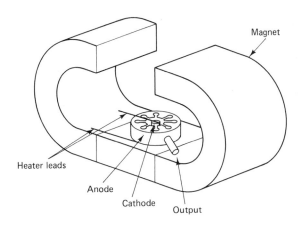

Figure 3-24 The magnetron.

When no magnetic field is present, the heated cathode emits a uniform flow of electrons (thermionic emission) to the plate or anode of our magnetron diode. However, if we were to mount a magnet around the diode and increase the magnetic flux, the electrons would actually proceed to the anode in a curved rather than a direct path. See Fig. 3-25.

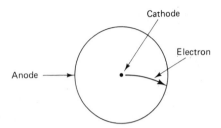

Figure 3-25   Electron path to anode.

In Fig. 3-26 the magnetic field has been increased further to cause the electrons to bend and just miss the anode in a circular orbit. This desired amount of magnetic flux is known as the *critical value.*

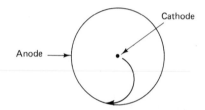

Figure 3-26   Electron skims past anode.

Just for the sake of discussion and to further illustrate the principle of magnetic flux influencing electrons, Fig. 3-27 shows the magnetic field increased beyond the ideal critical value, causing the electrons to travel in a circular orbit that is now too small.

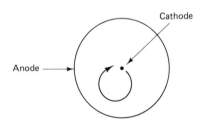

Figure 3-27   Electron travels in a smaller orbit.

The anode block is in fact made up of a number of resonant cavities, all of the same size and therefore tuned to the same frequency. Each slot and cavity is equivalent to a value of capacitance and inductance, as seen in Fig. 3-28. If a beam of electrons were made to pass the openings of a ring of resonant cavities, energy would be delivered to all the cavities and

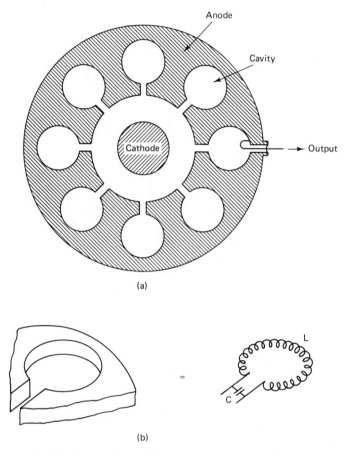

Figure 3-28 (a) Magnetron's cathode, anode, and output coupling; (b) a cavity has both capacitance (C) and inductance (L).

they would resonate at their frequency of oscillation. Let us examine this point a little more closely. The anode is at a positive potential, with respect to the cathode, and causes the electrons to accelerate in their curved path toward the cavities. Oscillations within the anode block of cavities move out into the interaction space of the tube, where these oscillations interact with the electrons travelling toward the cavities. The oscillations actually cause the electrons to accelerate and decelerate so that they arrive at the cavity at the same time the positive part of the oscillation is occurring, giving that oscillation the additional push needed to maintain ringing action.

Figure 3-29 illustrates a small hook in one of the cavities. This hook acts as a pick-up loop, extracting the microwave energy from this cavity (and thereby all others) when the magnetron oscillates.

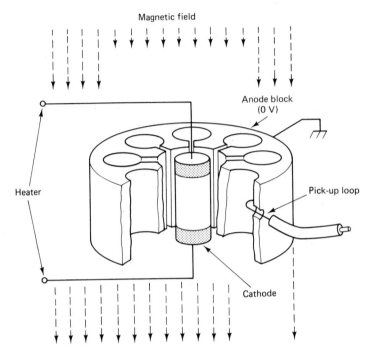

Figure 3-29   Magnetron output coupling.

The two largest applications for these high power (kW to MW) magnetron oscillators are in microwave ovens, where they supply continuous waves for fast heating of food, and in radar equipment for supplying very high-powered, short-duration microwave frequency outputs.

The two-cavity klystron.   The word *klystron* is derived from the Greek word *klyzein,* which means "the breaking of waves on the beach," a comparison that will become apparent after this explanation of the device.

The structure of a two-cavity klystron is illustrated in Fig. 3–30. A thermionic cathode emits a stream of electrons that immediately are attracted and travel to the large positive potential on the collector via the two cavities called a *buncher* and a *catcher.*

An input signal applied to the buncher causes oscillations within that cavity. These oscillations cause a speeding up and slowing down of the electron stream travelling from cathode to anode. Let us use a simple analogy to explain the effect that this would have. If a few cars at different points along a freeway were to slow down and others were to speed up, we would

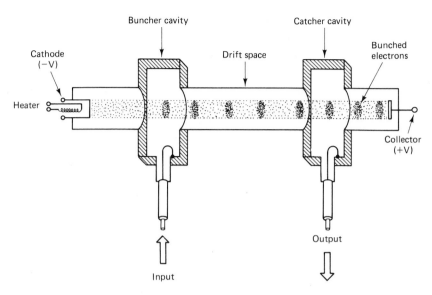

Figure 3-30   The two-cavity klystron.

end up with the fast cars catching up to the slow-moving vehicles, leaving spaces and causing bunches to occur on the freeway. Our buncher cavity oscillations cause bunches of electrons at the frequency of the input signal. The velocity-modulated electron bunches move away from the buncher cavity and now enter a drift space. The drift space, in fact, is where the accelerated electrons tend to catch up with the retarded electrons and form bunches.

Bunches of electrons pass the gap of the catcher resonator at a rate of one bunch per cycle of input signal. These bunches induce oscillations in the catcher resonator, and these oscillations are transferred out of the cavity by using loop coupling. The bunches, after exciting the catcher cavity, finally reach the collector in bursts or waves, which explains the meaning of the Greek word *klyzein*.

A small input signal has been used to velocity-modulate a large electron beam so that the output probe at the catcher will couple out an amplified version of the input signal. Here the two-cavity klystron was utilized as a microwave amplifier. Any amplifier, as you know, can be made to oscillate by feeding some of the amplified output, in our case from the catcher, back to the input at the buncher. This is known as *positive* or *regenerative feedback*. Figure 3-31 illustrates the two-cavity klystron with an external coaxial feedback connection.

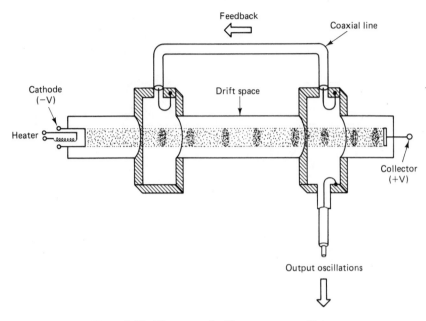

Figure 3-31 The two-cavity klystron as an oscillator.

The reflex klystron. The reflex klystron, a low-power oscillator, is illustrated in Fig. 3-32. It has only one resonant cavity, and unlike many oscillators it does not require external feedback; instead, it provides its own internal positive feedback. The internal feedback is achieved by replacing the positively biased collector with a negatively biased repeller.

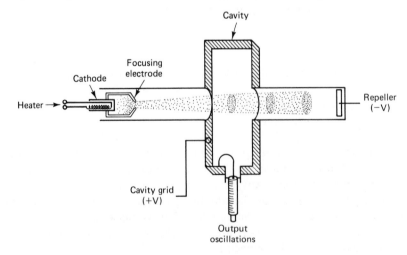

Figure 3-32 The reflex klystron oscillator.

In order to understand the operation of the reflex klystron oscillator, we must assume, as we would with any other oscillator, that oscillations have started, and must proceed to explain how oscillations can be maintained.

Let us now take a closer look at the operation of this microwave device. A thermionic cathode releases a stream of electrons that are accelerated towards a positive DC voltage present on the cavity grid. As the electrons pass through a gap in the cavity toward the repeller, the oscillations within the cavity will vary the speed of the electrons and set up bunches that drift toward the repeller. As they approach the large negative voltage of the repeller, the bunches of electrons will begin to slow down, stop, and eventually start to travel in the reverse direction back towards the cathode, and consequently will reenter the cavity for the second time. Upon entering the cavity the bunched electrons (one bunch per cycle of oscillation), supply a kick or push, which is needed to sustain oscillations within the cavity. The bunched electrons that are repelled back to the cavity are the reflex klystron's internal positive feedback system.

Figure 3-33 shows a reflex klystron with a waveguide output mount. The frequency of oscillation can be mechanically changed a large amount (coarse tuning) by adjusting the tuning screw and physically changing the size of the cavity and therefore the frequency of the oscillation.

The frequency can be electronically changed, but only a small amount (fine tuning), by varying the repeller voltage, which changes the transit time

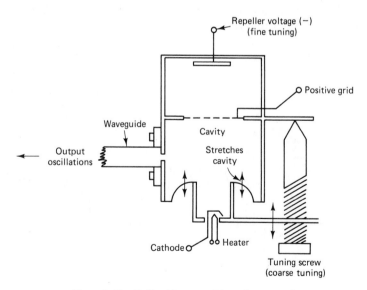

Figure 3-33 Reflex klystron with tuning control.

between first and second passage through the cavity. This will change the rate at which our bunches of electrons return to excite the cavity and maintain oscillations; by changing this rate you will change the oscillating frequency.

Reflex klystrons are used as low power (mW to kW) local oscillators in microwave superheterodyne receivers and as a signal source for general laboratory work.

Traveling wave tube (TWT). Figure 3–34 shows the internal structure of the tube. Once again a thermionic cathode emits electrons which are accelerated toward a large positive potential on the collector, but unlike other microwave amplifiers this tube does not house a resonant cavity. The signal to be amplified is applied at the waveguide input and travels along the helix, which can be thought of as a large inductor within the tube. The coil is in fact quite long and at microwave frequencies our wavelength, as we already know, is very small, so many RF cycles occur within the length of the helix. These cycles set up electromagnetic fields that interact with the electron stream travelling from cathode to anode. At points where RF voltage is positive, nearby electrons are accelerated; where the RF voltage is negative, the electrons are made to slow down. Thus, once again, bunches of electrons are produced all along the tube within the helix. As these bunches of electrons move towards the collector, their charges induce voltages in the helix that add to the electromagnetic waves of the RF signal input. This interaction takes place along the whole length of the tube, causing continuous interaction and therefore amplification. Consequently, it is no surprise to find out that the gain of a TWT is proportional to the length of the tube, or, more important, the length of the helix.

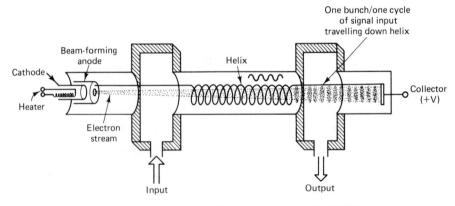

Figure 3–34 The travelling wave tube amplifier (TWT).

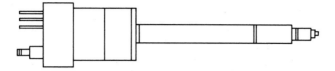

Figure 3-35   A travelling wave tube's appearance.

Figure 3-35 shows the external appearance of a TWT, approximately fourteen inches in length.

MICROWAVE DIODES

Point-contact diodes.   These were the first solid state devices developed in 1940 for radar. The diode consists of a semiconductor against which the end of a fine wire, known as a *catwhisker,* is pressed; see Fig. 3-36.

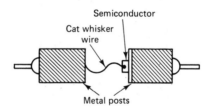

Figure 3-36   The point-contact diode.

The characteristics of the point-contact diode are quite different from those of the regular junction diode. The point-contact diode has the advantage of having very low capacitance affect across the two junctions of the diode. A cutaway view and schematic symbol are shown in Fig. 3-37.

The biggest application for the point-contact diode was in the mixer stage of a radar's superheterodyne receiver. As we mentioned previously, it is hard to encourage microwave energy to travel down transmission lines. It is also very difficult to send a radar signal through ten or twelve stages of amplification. Radar transmitters radiate power in kilowatts; receivers pick up reflections from targets that are in microwatts. These very small reflections or echoes will need to be amplified by ten stages of amplification at least to make them a usable signal that can be processed to find the targets' range and bearing. The conventional amplifiers are most suitable for this operation, but they will not amplify microwave signals. A mixer is used in the receiver of a radar system to mix the incoming signal or target information sitting on the high microwave frequency with a continuous-

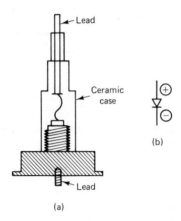

(b)

(a)

Figure 3-37   Point-contact diode. (a) Appearance; (b) schematic symbol.

wave low-power microwave oscillation. The mixer produces the sum and difference of the two input signals, that is, the local oscillator and the echo signals. Both the sum and difference contain the signal or echo information. The difference is our choice, as it is now a low frequency with the echo information sitting on top of it, which can be amplified by conventional amplifiers.

Figure 3-38 illustrates this balanced or hybrid mixer. The crystals or

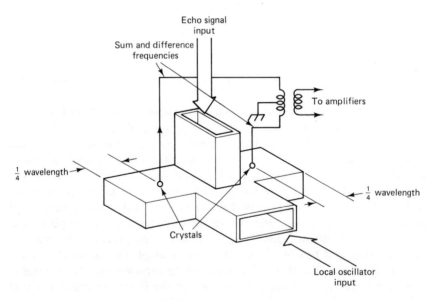

Figure 3-38   The balanced or hybrid mixer.

point-contact diodes are placed into the waveguide, one quarter-wavelength away from the short-circuited waveguide ends, as shown. The local oscillator signal and echo signal are both applied into this mixer at the same time. There will be a condition when the two signals applied are in phase at crystal one, and out of phase at crystal two, and vice versa. The end result will be that both the sum and difference of the two inputs will be pulled out of the mixer and coupled via the transformer to the input of the first-stage amplifier. There the amplifier will select the difference by tuning.

The Schottky diode. This diode is also known as the *hot carrier* or *barrier diode* and performs the same operation as the point-contact silicon diode just mentioned. The Schottky diode has completely replaced the fragile point-contact diode because it is more robust and reliable and has a higher current rating.

The Gunn diode. Figure 3–39 shows the physical appearance of a Gunn diode. It can be used for supplying continuous-wave oscillations ranging from 6 to 18 GHz.

A DC bias voltage is applied across the diode; when this bias is taken below a certain threshold level, the mobility of the electrons at the cathode is reduced. These lower-mobility electrons cause a bunching effect as the electrons ahead move on, leaving a depletion zone, and those behind catch up. This bunch is discharged as a large current pulse, and a new bunch starts at the cathode. The rate at which the current pulses occur is determined by the distance or thickness between cathode and anode.

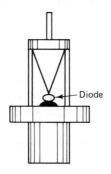

Figure 3–39 The Gunn diode.

Let us now mount the diode in a resonant cavity, as seen in Fig. 3–40. The bunches of electrons ejected out of the diode into the cavity will cause and maintain oscillations. These oscillations can be coupled out by the loop coupler seen in Fig. 3–40. This oscillator is tunable by making one

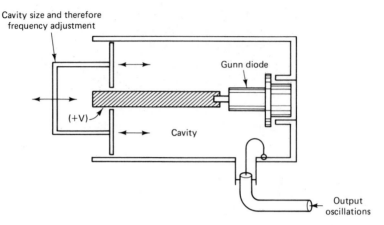

Figure 3-40   The Gunn diode oscillator.

end wall adjustable and therefore change the cavity size and frequency of oscillation.

The advantage of using a Gunn diode local oscillator instead of a reflex klystron is that the Gunn diode requires a DC bias of approximately 7 volts, whereas a reflex klystron requires a repeller voltage of approximately -800 volts, and therefore a high-voltage power supply.

DETECTING MICROWAVE ENERGY IN WAVEGUIDES

Solid state diodes, such as the point contact or Schottky diode, and bolometers are inserted in a waveguide section as illustrated in Figs. 3-41(a) and 3-41(b).

The diodes in the waveguide pick up the microwave energy, rectify it, and then feed the DC relative power indication to operate a DC meter. They can also detect modulation on a microwave carrier, and view this modulation on an oscilloscope.

A device that changes its resistance when heated is called a *bolometer*. There are two different types:

1. Barretters, which have a positive temperature coefficient.
2. Thermistors, which have a negative temperature coefficient.

The temperature coefficient, whether negative or positive, means that the resistance of our bolometer will change from its original value, as the

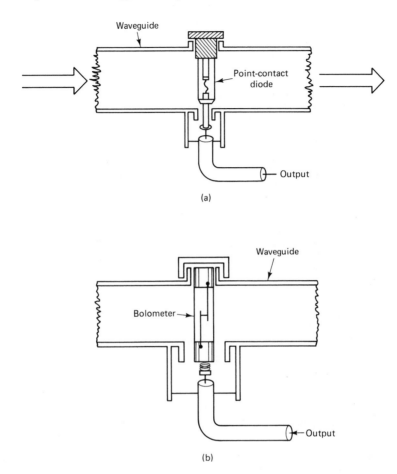

Figure 3-41   (a) Detector and mount; (b) bolometer and mount.

temperature increases. That is, thermistors have a negative temperature coefficient, which means that as temperature increases due to an increase in microwave energy passing through the waveguide, the thermistor element will heat up and its resistance will go down.

The thermistor could be placed in the fourth arm of a bridge, as seen in Fig. 3-42, and the DC meter will measure the bridge unbalance and indicate a relative value of microwave energy; that is, the more microwave energy in the waveguide, the lower the thermistor's resistance, the greater the bridge imbalance, and therefore the greater the relative power indication on the meter.

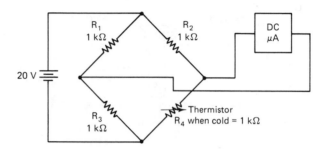

Figure 3–42 Thermistor in bridge.

SUMMARY

1. A hollow waveguide can carry microwave energy in an infinite number of ways called *modes,* and every mode is either a transverse electric (TE) or transverse magnetic (TM). In the transverse electric mode, the electric field is crossing the direction of propagation, whereas the transverse magnetic mode has the magnetic field crossing the direction of propagation arrow.

2. A waveguide is a high-pass filter cutting off frequencies below a value that is dependent on its size; this frequency is referred to as its *cutoff frequency;* the smaller the size of the waveguide, the higher its cutoff frequency.

3. Waveguides are normally made of metal; however, they must be handled with great care to avoid denting or scratching of the interior. Any bump will cause a reflection of our microwave energy, which will reduce the power of our travelling wave.

4. Resonant cavities replace LC circuits at microwave frequencies. The smaller the resonant cavity, the smaller the capacitive and inductive value, and so the higher its resonant frequency.

5. Apertures and probes are used to input and output microwave energy.

6. A magnetron is a high-power vacuum tube diode oscillator, which utilizes large magnets and resonant cavities.

7. A two-cavity klystron can be used to amplify microwave signals or to produce low-power continuous oscillation. The reflex klystron, which has only one cavity, is a low-power oscillator.

8. The travelling wave tube uses velocity modulation to achieve amplification of a microwave signal.

9. Some of the semiconductor diodes discussed included the point-contact, the Schottky, and the Gunn diodes.

REVIEW QUESTIONS

1. The dominant mode for rectangular waveguides is
   (a) TE_{11}
   (b) TM_{11}
   (c) TME_{101}
   (d) TE_{10}
2. A travelling wave tube amplifier:
   (a) Uses magnets to bend electrons
   (b) Has a repeller
   (c) Has buncher and catcher cavities
   (d) Contains a helix
3. The most efficient type of transmission line for a microwave system is
   (a) Twin lead
   (b) Coaxial cable
   (c) Two-wire transmission line
   (d) All of the above
   (e) Waveguide
4. In a waveguide system, travelling waves are more desirable than standing waves. True or false?
5. The frequency determining proportion of a magnetron is most similar to
   (a) A resonant cavity
   (b) A shorted line
   (c) An open line
   (d) A cathode
6. As microwave frequencies increase, the physical size of the waveguide used will
   (a) Decrease
   (b) Remain the same
   (c) Increase
   (d) None of the above
7. Energy can travel down the interior of a rectangular waveguide by means of:
   (a) Transverse electric mode
   (b) Transverse resistive mode
   (c) Transverse magnetic mode
   (d) Both (a) and (c)
   (e) Both (a) and (b)

8. Gunn diode oscillators operate on the principle of:
   - **(a)** An electron bending between cathode and anode
   - **(b)** A large repeller voltage to send electrons back to the cathode
   - **(c)** A helix
   - **(d)** Electrons bunching and pulsing current into a resonant cavity
   - **(e)** All of the above

9. The primary application of crystal rectifiers in microwaves is as a
   - **(a)** Detector or mixer
   - **(b)** High-power detector
   - **(c)** Modulator
   - **(d)** Plate supply rectifier
   - **(e)** None of the above

10. The space between cavities in a two-cavity klystron is called:
    - **(a)** Anodes
    - **(b)** Bunching space
    - **(c)** Cathode space
    - **(d)** Drift space

11. At microwave frequencies, we use:
    - **(a)** Waveguides
    - **(b)** Klystrons
    - **(c)** Cavity resonators
    - **(d)** All of the above
    - **(e)** None of the above

12. What is the purpose of dummy loads?

13. Give the names of two dummy loads:
    - **(a)** _____
    - **(b)** _____

14. Give the names of two variable attenuators:
    - **(a)** _____
    - **(b)** _____

15. A travelling wave tube can be made to amplify or oscillate. True or false?

16. Describe what is meant by the cutoff frequency of a waveguide.

17. A waveguide acts as a _____ pass filter.

18. Illustrate a transverse electric mode and transverse magnetic mode pattern and explain the difference between the two.

19. What are the names of the two cavities in a two-cavity klystron?
    - **(a)** _____
    - **(b)** _____

20. Illustrate and explain the operation of a two-cavity klystron oscillator.

21. What are the two advantages a reflex klystron has over a two-cavity klystron oscillator?
    - **(a)** _____
    - **(b)** _____

22. Describe what is meant by *thermionic cathode.*

23. The gain of a TWT is proportional to:
    (a) The length of the tube
    (b) The type of input and output coupling
    (c) None of the above
    (d) Both (a) and (b)

24. Illustrate and explain the operation of a balanced or hybrid mixer.

25. What are the two different types of bolometers?
    (a) _____
    (b) _____

26. Describe what is meant by:
    (a) A positive temperature coefficient
    (b) A negative temperature coefficient

27. A thermistor has:
    (a) A positive temperature coefficient
    (b) A negative temperature coefficient
    (c) No change is resistance
    (d) None of the above

28. Describe briefly what is meant by bunching effect in relation to a Gunn diode.

29. A barretter has:
    (a) A positive temperature coefficient
    (b) A negative temperature coefficient
    (c) No change in resistance
    (d) None of the above

30. Frequency of resonance $= \dfrac{1}{2\pi \sqrt{LC}}$

    The smaller the resonant cavity, the _____ the values of capacitance and inductance, and consequently the _____ the frequency of resonance.
    (a) Larger, higher
    (b) Smaller, higher
    (c) Larger, lower
    (d) Smaller, lower

4

MICROWAVE ANTENNAS

Objectives

After completing this chapter, you will be able to:

1. State how the gain of an antenna is calculated.
2. Define:
   a. Decibels
   b. Beamwidth
   c. Bearing resolution
3. Explain why a vertical lobe structure can become broken up and distorted.
4. List the three types of horn radiators.
5. Identify the four types of parabolic reflectors.
6. State the difference between front- and rear-fed paraboloids.
7. Describe the cassegrain feed system.
8. Explain the function and operation of the slotted waveguide antenna.

An antenna when used at microwave or lower frequencies is just a metal conductor that converts the transmitter output signal current to electromagnetic waves. These waves are propagated to the receiving antenna, where they induce currents that are fed as an input to the receiver.

GAIN OF AN ANTENNA

When one is trying to calculate the gain of an antenna, there is obviously going to be a large difference between one and another because of all the different types. We must have a way of comparing different antennas, and the easiest way to achieve this is to compare them all to a known fixed reference point. An *isotropic antenna* is one that radiates equally in all directions, and can be used as a reference to compare the effectiveness of all our directional antennas.

Figure 4-1 shows an isotropic antenna being fed with 1 kilowatt of power and radiating this power in all directions. One mile away from the transmitter the field strength is equal to 1 milliwatt.

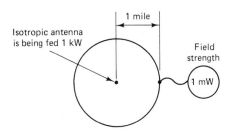

Figure 4-1 An isotropic antenna.

Our directional antenna, shown in Fig. 4-2, concentrates the electromagnetic waves in one direction, and to achieve the same field strength of 1 milliwatt one mile away requires only 1 watt of power to be supplied to this antenna.

The formula for calculating the gain of a directional antenna is

$$\text{Gain} = \frac{\text{power to isotropic antenna}}{\text{power to directional antenna}}$$

(for equal field strength)

So the gain of our directional antenna from the example just mentioned would be

$$\text{Gain} = \frac{\text{power to isotropic antenna}}{\text{power to directional antenna}} = \frac{1\text{KW}}{1\text{W}} = 1000$$

This value can be compared with other values given by other directional antennas, and the higher the result, the better the directional properties of the antenna. However, gain (designated G) is measured in decibels.

The decibel, abbreviated db, is the standard unit for expressing the amount of gain or loss in transmission power levels, and also is used as a way to compare the difference between two power levels at different points. If the power intensity increases, the value is expressed as a positive number of dbs; a decrease is represented as a negative value of dbs. If we were to have no increase in intensity, the power change would be represented as 0 db.

To calculate how much we have gained with our antenna, we must consider how much power we apply to the antenna and how much power we received out of that same antenna and convert this to a decibel value.

$$\text{Gain of antenna (db)} = \frac{10 \times \log \text{power out}}{\text{power in}}$$

Example

Power in = 50 watts   Power out = 500 watts

What is the gain of this directional antenna in decibels?

$$\text{Gain of antenna} = \frac{10 \times \log \text{power out}}{\text{power in}}$$

$$= 10 \times \log \frac{500 \text{ watts}}{50 \text{ watts}}$$

$$= 10 \times \log \times 10 = \underline{10 \text{ db}}$$

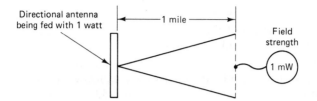

Figure 4-2   A directional antenna.

As another example, a directional transmitting antenna has a gain of +30 db. This means that the antenna will output a signal +30 decibels greater or 1000 times greater than that from an isotropic antenna which is fed from the same transmitter. The power ratio in this case is equal to 1000.

Table 4-1 lists common decibel values for power ratios.

Let us now place a receiving antenna in a position so that it will intercept a microwave beam. The gain the antenna achieves is proportional to the size or area of the antenna; that is, the larger your antenna, the more of the transmitted beam can be captured and passed on to the receiver. Furthermore, the gain is going to be inversely proportional to the wavelength of the receiver signal; that is, as wavelength decreases, gain increases. To further explain the last sentence we must first consider that if wavelength has decreased, then frequency would have increased; if frequency had increased, our receiving antenna would now be capturing more cycles and therefore more energy in a given unit of time.

$$G = \frac{4\pi \times A \times \eta}{\lambda^2}$$

η = aperture efficiency λ = wavelength in cm
G = gain in decibels A = aperture area

TABLE 4-1

Power Ratio	Decibels
1000	30
100	20
10	10
2	3
1	0
1/2	−3
1/10	−10
1/100	−20
1/1000	−30

The previous formula proves what has just been discussed—that the larger the antenna's area and the smaller the wavelength (the greater the frequency) the better the gain of an antenna.

Similarly, the greater the size of a transmitting antenna, the better the directivity of that antenna; the higher the frequency, the more energy transmitted in a given unit of time. These both add up to a higher-gain transmitting antenna.

BEAM PATTERN

A beam antenna is one that concentrates its radiation into a narrow beam of definite direction, as illustrated in Fig. 4-3.

The sidelobes are a portion of the beam, other than the main lobe, in the horizontal plain; they are undesirable, as power is being wasted to either side instead of concentrating that energy into the main beam. It is therefore

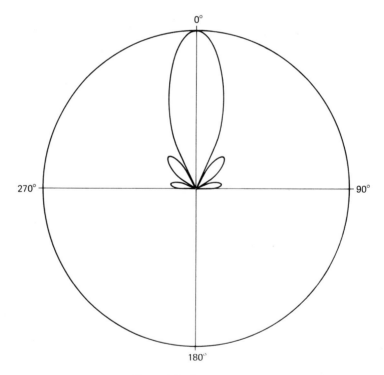

Figure 4-3 Beam pattern.

desirable when one is choosing an antenna to select one whose sidelobes are as small as possible. Also shown in the beam pattern diagram in Fig. 4-3 is the relative power decreasing (-db), from the maximum point to zero.

HORIZONTAL BEAMWIDTH

Beamwidth is defined as the angular width of a beam measured between the lines of half-power intensity. The angle subtended between these two lines is illustrated in Fig. 4-4.

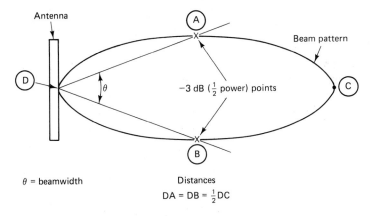

Figure 4-4   A directional antenna's beamwidth.

BEARING RESOLUTION

Bearing resolution is the ability of a radar to distinguish between two targets at the same range but on slightly different bearings, as illustrated in Fig. 4-5. Bearing resolution depends on the horizontal beamwidth of the antenna; the narrower the beam, the better a radar can discriminate between the two targets. To fully understand this point, we must talk a little further on a few radar principles.

The horizontal beam from the ship's antenna is highly directional and rotates through 360 degrees in approximately three seconds. When the beam encounters a target, energy is returned back to the ship, where it is displayed on a screen to indicate both the ship's range and bearing. Consequently, if our beam were to strike two targets at slightly different bearings, we would receive two targets, but they would be blurred together on the screen. It is

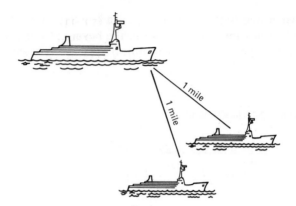

Figure 4-5   Bearing resolution.

therefore most desirable that the horizontal beamwidth be as small as possible so that the beam can fit between these two targets in bearing, and therefore can return two separate signals and so produce two distinguishable target indications on the screen.

Let us take an example and illustrate it to further simplify this concept. Figure 4-6 shows a beam pattern whose right-hand edge is hitting target 1. An echo, or return signal, from this target will show up on our radar screen as illustrated.

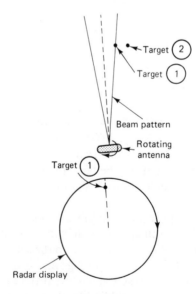

Figure 4-6

Figure 4-7 shows that our beam pattern has now rotated a small amount so that now target 1 is directly in the beam. The return of energy will show up on the display screen as illustrated.

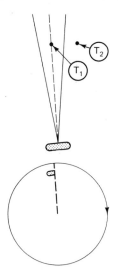

Figure 4-7

Figure 4-8 shows that the beam pattern has rotated again further, and now targets 1 and 2 are in its path. Two returns from both targets will

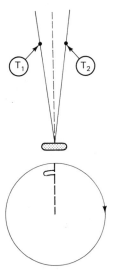

Figure 4-8

result, but the display will not be able to distinguish between the two and will therefore combine them both as one target.

Figure 4–9 and Fig. 4–10 show the beam pattern still rotating in a clockwise direction until it moves away from target 2.

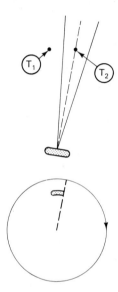

Figure 4–9

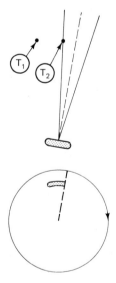

Figure 4–10

To summarize these illustrations, we can say that if two or more objects are contained within the beamwidth, a continuous echo will be painted on our screen—but if our beamwidth is narrow enough to fit between two targets, then two echoes and two separate paints on our screen will result.

VERTICAL LOBE STRUCTURE

As we discussed in Chapter 2, only the direct and reflected waves are present when one is operating at microwave frequencies. This, however, causes a problem, as will now be explained.

Figure 4-11 shows the ideal vertical plane coverage. Figure 4-12 shows the typical vertical plane coverage for a ship's marine navigation radar.

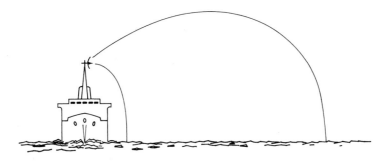

Figure 4-11   Ideal vertical plane coverage.

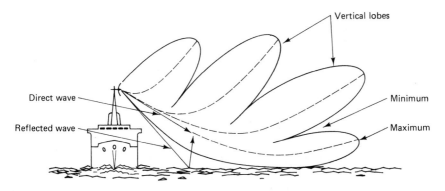

Maximum — when direct and reflected wave were in phase
Minimum — when direct and reflected wave were out of phase

Figure 4-12   Actual vertical plane coverage.

Due to reflections against the sea surface, the vertical lobe pattern is broken up and distorted. The radiated energy in the vertical plane is now the vector sum of the directly transmitted and reflected energy from the sea surface; that is, the reflections from the sea surface may at one point arrive in phase with the direct wave, causing a maximum, or these reflections may arrive out of phase with the direct wave, causing a minimum. Targets some distance away may fall between the lobes and cannot be detected until they move into a lobe where the two waves have combined to cause a maximum. Very low-lying targets can pass underneath the first lobe and not be detected.

The approximate number of lobes is proportional to the antenna's height, and can be calculated by using the following formula:

$$\text{Number of lobes} = \frac{\text{antenna height (in meters)}}{\lambda \text{ (wavelength in cm)}}$$

Example

$$\text{Antenna height} = 15 \text{ meters} \qquad \text{Wavelength} = 3 \text{ cm}$$

$$\text{Number of lobes} = \frac{\text{antenna height}}{\text{wavelength}} = \frac{15}{3} = 500 \text{ lobes}$$

EFFECT OF SHIP'S ROLL ON RADAR

Vertical beamwidth, unlike horizontal beamwidth, is much larger for ships' radar systems. See Fig. 4-13.

The vertical beamwidth of the antenna should be such to ensure that targets at maximum and minimum range still remain in the beam coverage even when the ship is rolling quite badly.

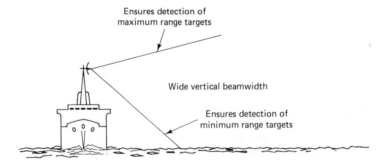

Figure 4-13   Vertical beamwidth.

Typical radar antennas will therefore have a very narrow horizontal beamwidth for good bearing resolution, and a large vertical beamwidth to ensure no loss of maximum and minimum range targets in large swells.

ANTENNA TYPES

Generally speaking, a transmitting antenna consists of a feed system to supply the microwave energy and a focusing system to direct the energy into a beam. For reception of microwave signals, the focusing system collects and focuses the energy to the feed system, which transfers this microwave information to a receiver.

WAVEGUIDE TO FREE-SPACE MATCHING

Before starting this subject, let us first review a word that frequently occurs with transmission lines and antennas—impedance.

Impedance is the total opposition to the flow of alternating current. If we wish to transfer energy from one point to another, we must have a condition known as an *impedance match;* that is, source impedance is equal to load impedance. Impedance matching ensures that maximum power is transferred from your source to the load with no reflections occurring and therefore no loss of power between the two.

Waveguides have an impedance that is generally dependent on the waveguide size, while the impedance of free space is the same as a waveguide of infinite height and width.

The three horn radiators seen in Fig. 4-14 are just open pieces of waveguide that flare out. The flare ensures the impedance match between waveguide and free space, and also has the advantage of being directional.

(a)   (b)   (c)

Figure 4-14   Horn Radiators. (a) Rectangular; (b) pyramidal; (c) conical.

PARABOLOID ANTENNA

These antennas are extensively used in earth and satellite point-to-point communication systems. A radiation beam similar to a searchlight is produced by a surface called a *paraboloid*. See Fig. 4-15, which shows some examples of paraboloid antennas used for receiving television transmissions from satellites.

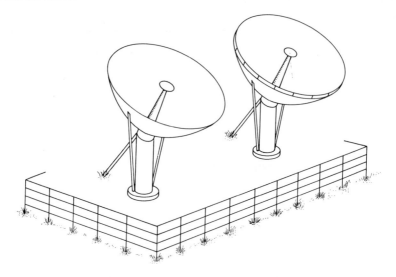

Figure 4-15   Satellite TV receiving antennas.

Every parabola has an axis, which is at an angle of 90 degrees to the directrix, as seen in Fig. 4-16. The parabolic reflector, commonly called a

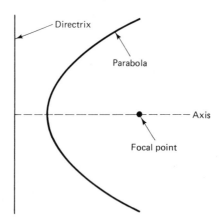

Figure 4-16   The parabola.

dish, is fed from a point source, typically a horn radiator, located at the focus point.

Figure 4-17 illustrates the optical design principle for paraboloids. Since microwave frequencies behave similarly to light waves, it is possible to design a reflector by applying the optical law that states, "Angle of incidence is equal to angle of reflection."

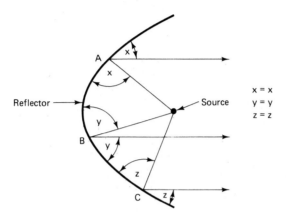

Figure 4-17   Optical design of paraboloids.

This diagram illustrates that incident microwave energy from a source feed falling on point *A* will be reflected in the desired direction, and the two angles *X* will be equal. This statement will be true for points *B* and *C* and angles *Y* and *Z,* respectively.

Figure 4-18 shows four forms of parabolic reflectors. Let us now proceed to discuss each one of these antenna shapes separately:

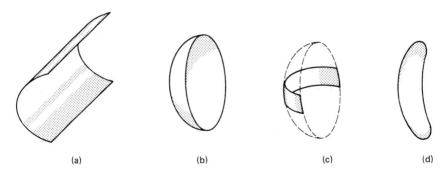

Figure 4-18   Parabolic reflectors. (a) Parabolic cylinder; (b) paraboloid; (c) truncated paraboloid; (d) orange peel paraboloid.

1. The first reflector is in the form of a parabolic cylinder which will be fed from a line source placed along the focal line, as seen in Fig. 4–19.

A cylinder of this type is about twenty feet in length, projects a beam with a horizontal beamwidth of approximately 10 degrees, and has a vertical beamwidth of about 3 to 6 degrees.

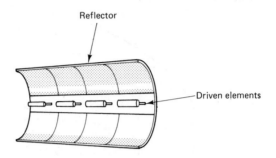

Figure 4-19 A parabolic cylinder with feed.

2. The paraboloid that was illustrated in Fig. 4–18(b) could be subdivided into two classes:

(a) Front feed

(b) Rear feed

Figure 4–20 illustrates these two alternatives, which will now be covered in greater detail.

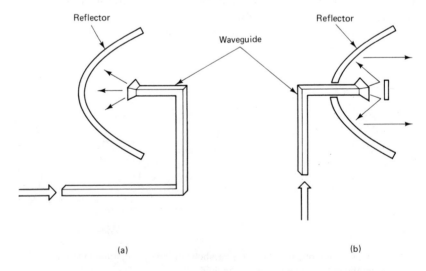

(a)   (b)

Figure 4-20 Front- and rear-fed paraboloids. (a) Front feed; (b) rear feed.

(a) In the front-feed illustration, the feed waveguide is brought around the edge of the antenna to the focal point where the energy is radiated, typically a tapered horn. Figure 4-21 shows a front-fed system using a tapered horn to radiate energy to a parabolic reflector.

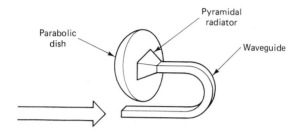

Figure 4-21   Front-fed parabolic reflector with pyramidal horn feed.

(b) In the more popular rear-fed approach, the waveguide comes through the center of the dish and radiates to the paraboloid from the focal point. One method to rear-feed a paraboloid is by use of a dipole, as shown in Fig. 4-22.

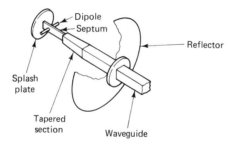

Figure 4-22   Rear feed system.

The dipole, or half-wavelength antenna, is mounted one-quarter of a wavelength away from the splash plate or subreflector. The function of the splash plate is to reflect energy from the dipole back towards the main reflector. This is illustrated in Fig. 4-23.

The quarter-wavelength gap between dipole and splash plate is arranged so that reflected waves have to travel one-half wavelength before returning and meeting up with the original transmitted wave from the dipole. Due to reversal of phase on reflection from the splash plate surface plus the half-wavelength distance, the reflected wave from the splash plate and the direct wave from the dipole move toward the main reflector in phase.

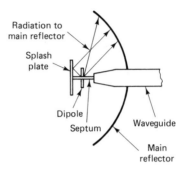

Figure 4-23   Splash plate principle.

Cassegrain feed.   Another rear-feed method is the cassegrain feed, which was derived from astronomical reflecting telescopes using the same principle and named after an eighteenth-century astronomer.

Figure 4-24 shows a hyperboloid (secondary reflector) placed at the focal point of the paraboloid (primary reflector). The microwave energy is radiated from the feed horn and then is reflected by the hyperboloid to the paraboloid, which collimates (makes parallel) the rays.

Almost all large satellite communication earth stations are cassegrain-equipped, and only the consumer-type dishes are still focal-feed.

With focal-feed antennas a longer waveguide run is required because the microwave signal must be guided by waveguide to the focal point,

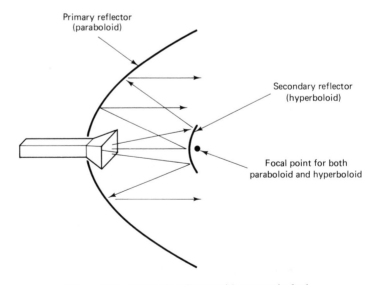

Figure 4-24   Parabolic reflector with cassegrain feed.

whereas with cassegrain feed the waveguide ends at the center of the paraboloid. The shorter the transmission line, the lower the noise, and consequently the higher the antenna's gain. Figure 4–25 shows a typical cassegrain-fed paraboloid reflector.

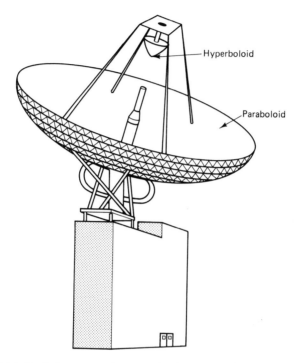

Figure 4–25 Typical overseas telecommunication earth station antenna with cassegrain feed.

3. Referring back to Fig. 4–18, let us now take a short look at the truncated paraboloid reflector, which is a paraboloid with its top and bottom section cut away.

In ships' marine navigational radar systems, it is desirable that the horizontal beamwidth should be as narrow as possible, and the vertical beamwidth fairly large. The truncated paraboloid will achieve this desired condition because of its design. The sides of this antenna bend around and consequently concentrate the beam in the horizontal plane; however, the lack of a top and bottom means that there are no restrictions for the beam in the vertical plane. See Fig. 4–26, which illustrates the resultant beam produced by this antenna.

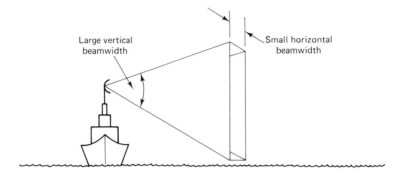

Figure 4-26 Typical radar's beam pattern (marine).

4. The orange peel paraboloid, so called because it resembles a section of orange peel, supplies the same beam characteristics as the truncated paraboloid just discussed. Figure 4-27 shows an orange peel paraboloid with a front horn feed.

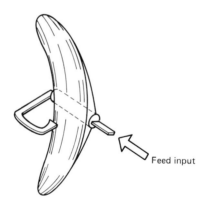

Figure 4-27 Orange peel reflector with front horn feed.

All the paraboloids just mentioned may be anywhere from approximately 3 to 30 meters in diameter, and in some cases the large reflectors are made in the form of wire mesh, which reduces weight and wind resistance. As long as the holes in the wire mesh are small compared to the wavelength of the signal being transmitted or received, the paraboloid acts as a solid surface with very little effect on the antenna's performance. See Fig. 4-28, which illustrates a wire mesh reflector being used for an IFF (identify friend or foe) system.

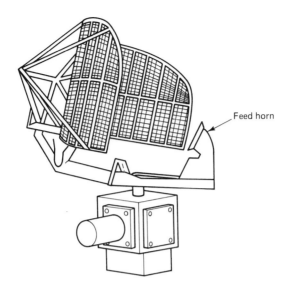

Figure 4-28   IFF wire mesh reflector.

THE SLOTTED WAVEGUIDE ANTENNA (SWG)

This antenna comprises a side slotted waveguide section, end fed and terminated in a matched load at the opposite end to the feed. This type of antenna is extensively used in marine navigational radar equipment. The narrow horizontal beamwidth and wide vertical beamwidth are formed by a flared horn that extends the whole length of the slotted waveguide section; a watertight fiberglass housing protects the antenna from environmental exposure. See Fig. 4-29.

Figure 4-30 illustrates the current distribution for the TE_{10} mode in a rectangular waveguide.

If a narrow slot is cut in the waveguide wall so that it does not disturb the current flow (at no angle), then the slot will be a nonradiating aperture. See Fig. 4-31.

If a slot is cut into the waveguide wall so that the current flow is disturbed, this aperture will radiate energy. Each slot will act as a single radiator. See Fig. 4-32.

If you refer back to Fig. 4-29, you can easily envisage that the power at the feed end will be greater than the power at the matched load end. If all the slots were the same, the beam pattern would not be symmetrical, as shown in Fig. 4-33.

To counteract and solve this problem, the size of the slots becomes

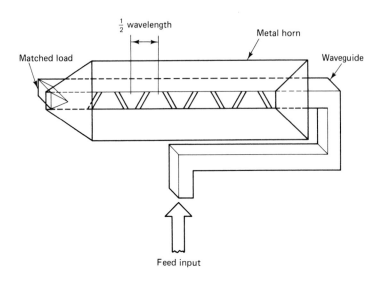

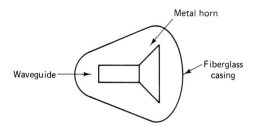

Figure 4-29   The slotted waveguide antenna.

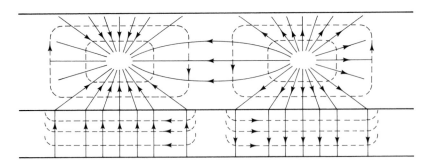

Figure 4-30   Current in rectangular waveguide.

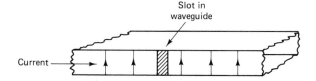

Figure 4-31   Nonradiating aperture.

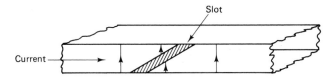

Figure 4-32   A radiating slot.

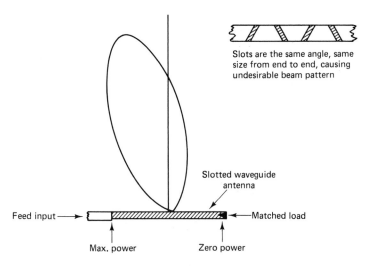

Figure 4-33   Unsymmetrical beam pattern.

larger and the angle of the slots increases. The larger the angle of the slot, the more they interfere with the current in the waveguide wall, and a larger disruption of current will cause an increase in the radiant energy as you move towards the matched load on the left-hand side of the diagram. The result is a symmetrical beam pattern, as shown in Fig. 4-34.

The advantages of the slotted waveguide antenna, as opposed to a paraboloid, is that it is light, compact, has less wind resistance, and side-lobes are greatly reduced because of the design, which is similar to a horn radiator, but a lot more efficient.

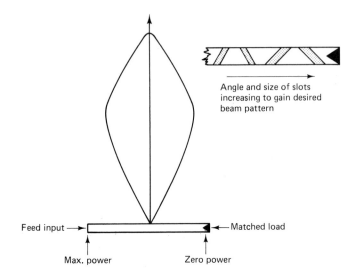

Angle and size of slots
increasing to gain desired
beam pattern

Feed input →   ← Matched load

Max. power   Zero power

Figure 4-34   Symmetrical beam pattern for SWG antenna.

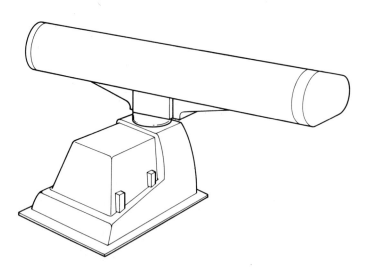

Figure 4-35   Ship's slotted waveguide radar antenna.

Figure 4-35 shows a navigational radar's slotted waveguide antenna on the scanner turning unit.

AIRBORNE RADAR ANTENNAS

Bombing, navigation, and search are some of the applications for aircraft radar equipment, which is normally mounted in a nonconducting airtight radome. The shape of this radome, wherever positioned, has been chosen to preserve aerodynamic design and not to impair the operation of the aircraft.

Airborne antennas are specially constructed to withstand vibrations, and the weight of the antennas and mechanisms needed to rotate these antennas must always be kept to a minimum.

The antenna will be mounted in a position best suited for the coverage desired; remember that an unobstructive view must always be maintained when using microwave frequencies. If only forward scanning is required, the antenna is mounted in the nose; if scanning the ground below and the horizon is required, the antenna may be mounted below the fuselage.

ANTENNA SAFETY PRECAUTIONS

One should always ensure that antennas and transmission lines are deenergized before even approaching their vicinity. Care should be taken to ensure that the equipment cannot be turned on when maintenance is being performed by removing the main fuses for the system, informing personnel, and placing indication on equipment to warn people of your situation.

Microwave antennas are always going to be high in order to gain maximum range. When working aloft, one should utilize safety harnesses at all times.

SUMMARY

1. An antenna is a metal conductor that converts the transmitter's output to electromagnetic waves. The receiving antenna captures these waves and feeds them to the receiver.

2. The gain of an antenna is basically equal to power out divided by power in. The value obtained is then converted to decibels, which are the standard unit for expressing the amount of gain or loss.

3. The horizontal beamwidth is the angular width of a beam in the horizontal plane measured between the lines of half-power.

4. Bearing resolution of a radar is directly proportional to the horizontal beamwidth of an antenna; that is, the better the horizontal beamwidth, the better the bearing resolution.

5. A transmitting antenna consists of a feed system to supply microwave energy and a focusing system to direct the energy into a beam.

6. A receiving antenna consists of a focusing system to collect the energy and focus it to a feed system, which transfers the energy to the receiver.

7. Horn radiators are used to match the impedance of a waveguide to the impedance of free space.

8. The paraboloid antenna is used in earth and satellite point-to-point communications. They can be front- or rear-fed, with the rear feed being most popular, as it provides the least amount of obstruction to our microwave energy.

9. A ship's radar antenna requires a large vertical beamwidth and a small horizontal beamwidth.

10. The slotted waveguide antenna is extensively used in marine radars, as it is light and compact.

REVIEW QUESTIONS

1. Calculate the gain of an antenna in decibels when

$$\text{Power out} = 1292 \text{ watts} \qquad \text{Power in} = 375 \text{ watts}$$

2. The decibel value for 1000 is
   (a) 20
   (b) -10
   (c) -40
   (d) 30

3. The power ratio for 20 db is
   (a) 0.5
   (b) 20

(c) 100

(d) 125

4. Calculate the number of vertical lobes when the antenna height is 30 meters, and we are transmitting a frequency of 40 Ghz.

5. Name the three types of horn radiators.

(a) _____

(b) _____

(c) _____

6. Name the four forms of parabolic reflectors.

(a) _____

(b) _____

(c) _____

(d) _____

7. A cassegrain feed antenna makes use of:

(a) A hyperboloid

(b) A paraboloid

(c) A rear feed

(d) All of the above

(e) None of the above

8. What is the function of the splash plate in the rear-fed paraboloid antenna?

9. The slotted waveguide antenna:

(a) Has slots cut in the waveguide wall to disturb current

(b) Has slots cut in the waveguide wall that do not disturb current

(c) Is a paraboloid

(d) All of the above

(e) None of the above

10. Calculate the gain of an antenna when the power to an isotropic antenna is equal to 1376 kw, and the power to our directional antenna for the same field strength is only 27.3 watts.

11. The bearing resolution of a ships radar is dependent on:

(a) The vertical beamwidth

(b) The horizontal beamwidth

(c) The scanner's height

(d) None of the above

12. With the aid of a diagram, illustrate the location of the parabola's focal point, axis, and directrix.

13. A parabolic reflector with cassegrain feed is

(a) Rear-fed

(b) Front-fed

(c) Side-fed

(d) None of the above

14. With a diagram, show and label all parts of a parabolic reflector with cassegrain feed.

15. A marine radar's antenna must provide a _____ vertical beam-width, and _____ horizontal beamwidth.
    (a) Small, large
    (b) Large, large
    (c) Small, small
    (d) Large, small

16. What advantage does a cassegrain feed paraboloid have over a focal-fed paraboloid?

17. Give two reasons why large reflectors are sometimes made of wire mesh.
    (a) _____
    (b) _____

18. Give two advantages the slotted waveguide radar antenna has over a parabolic radar reflector.
    (a) _____
    (b) _____

19. The slotted waveguide antenna is extensively used for:
    (a) Satellite TV reception
    (b) Voice communications
    (c) Marine radars
    (d) Microwave telephone links

20. With reference to Fig. 4-23, what is the distance between the dipole and the splash plate?

21. If a narrow slot is cut in the wall of a waveguide so that it disturbs current flow, that slot is said to be
    (a) A nonradiating slot
    (b) A radiating slot
    (c) None of the above
    (d) Both (a) and (b)

22. Referring to Fig. 4-8, explain why the vertical plane coverage for a ship's radar is broken up and distorted.

23. If a narrow slot is cut in the wall of a waveguide so that it does not disturb current flow, then that slot is said to be a
    (a) Nonradiating slot
    (b) Radiating slot
    (c) None of the above
    (d) Both (a) and (b)

24. As microwave frequencies act like light waves, reflectors are designed by applying the optical rule that "angle of _____ equals angle of _____."
    (a) Incidence, reflection
    (b) Refraction, reflection
    (c) Incidence, refraction
    (d) None of the above

25. SWG is an abbreviation for_____.

TWO MICROWAVE APPLICATIONS

Objectives

After completing this chapter, you will be able to:

SECTION ONE. RADAR

1. Explain the principle of radar.
2. Define the term two-way *travel time.*
3. Identify and explain the function of the four main units of a radar.

SECTION TWO. SATELLITE COMMUNICATIONS

1. Explain what is meant by a geostationary satellite.
2. Describe some of the satellites' uses.
3. List and explain the three main devices in a marine satellite communications system.
4. Explain a step-by-step communication procedure.
5. State the advantages of satellite communications.

The objective of this chapter is to give an introduction to two microwave systems, a "radar" and a "satellite communicator", both of which are covered in greater detail in the following two chapters.

PRINCIPLE OF RADAR

The word radar is an abbreviation for **ra**dio **d**etection **a**nd **r**anging, and that describes the principle of radar, which is to detect the direction (bearing) and range of anything that will reflect microwave energy.

Before radar, if a ship was caught in bad fog conditions near a coast line, the captain used to fire a shot or give a short blast on the ship's horn, and check how long it took for the returning echo. Since it was known that sound travelled at 1100 feet/second, if the echo returned after four seconds, the sound pulse must have taken two seconds to reach the cliff and two seconds to return; refer to Fig. 5-1.

These days ships transmit a short microwave burst of energy, and electronically calculate the time it takes before the echo is returned from the target to the ship; this time is directly proportional to the target's distance.

Microwaves, like radio waves, travel at 162,000 nautical miles/second or 300,000 kilometers per second.

If you got into your car and travelled at 20 miles per hour for one hour, it is easy to calculate that you would have travelled a distance of 20 miles. The formula we use is

Example Range = velocity (speed) × time
         = 20 mph × 1 hour
   Range = 20 miles

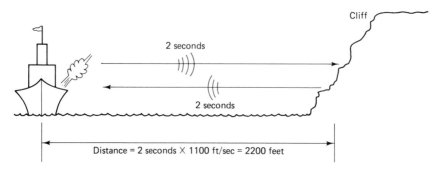

Figure 5-1

When considering microwaves, we know that velocity (speed) is fixed, so if we take a fixed range, for example, 1 nautical mile (nm), we will be able to calculate the time it takes for a microwave pulse to travel the distance of 1 nm.

If we transpose the original formula, we have

$$\text{Time} = \frac{\text{range}}{\text{velocity}} = \frac{1 \text{ nm}}{162{,}000 \text{nm/sec}} = 6.17 \ \mu s$$

Figure 5-2

But this, 6.17 μs, is only the one-way travel time; the pulse will have to return so that it can be detected by the ship. Therefore an echo will return from a target at one nautical mile (nm) after 2 × 6.17 μs = 12.34 μs.

For ease of calculation we generally quote the *two-way travel time for 1 nm as 12.5 μs.*

A BASIC RADAR BLOCK DIAGRAM

Refer to Figure 5-3, which is a block of the four main units, the transmitter, receiver, display unit, and antenna (Æ).

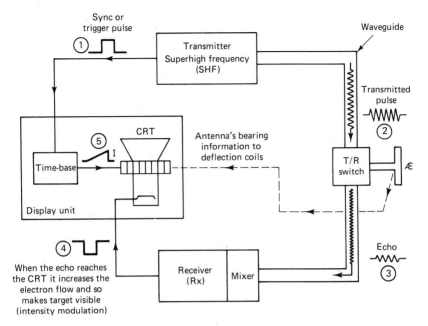

Figure 5–3

The Transmitter Block.   Houses a timing circuit which produces trigger pulses at regular intervals. This synchronization pulse causes two circuits to be activated:

1. The transmitter, which produces a high-powered, microwave frequency pulse which travels to the antenna via the T/R switch.
2. The time base circuit in the display unit, which is told to start producing a ramp waveform.

The Transmit/Receive Block (T/R).   Automatically connects the transmitter to the antenna for the duration of the microwave pulse and then switches, to connect the antenna to the receiver for the interval between pulses when we are waiting for returning echoes.

The Receiver.   This high-gain, low-noise, superheterodyne receiver amplifies the very small reflected pulses and presents them to the display.

Display Unit.   A cathode ray tube (CRT) is used as a plan position indicator (PPI), which means an electron beam from the cathode travels to the screen of the CRT and produces a spot in the center which is our position. Any target received will appear at a range and bearing relative to our position (the center point on the CRT); it therefore gives you a plan view indication of targets' positions.

In order to understand exactly how we display the echoes on the CRT, we will use an example.

Example

The antenna is pointing to 0 deg, and a deflection coil which is mounted round the neck of the CRT is positioned so that if any current is passed through the coil, a magnetic field will be developed; this magnetic field will deflect the electron beam from its natural center position on the CRT toward 0 deg. The amount of current determines the amount of deflection. These deflection coils, like the antenna, are free to rotate. In fact, the antenna feeds bearing information to the deflection coils, so if the antenna is pointing to 90 deg, the rotating deflection coils have been put in the same position. When current is applied to the coil, it will deflect the beam at the same angle the antenna is pointing, 90 deg.

If you refer now to the time-related waveforms (Fig. 5–4), we will be able to get a good understanding of what is happening and when.

As mentioned previously, the trigger circuit in the transmitter block starts the transmitted pulse and is fed to the display to start the time-base waveform. The time-base block, as its name implies, produces a timed waveform that will produce a base line on the CRT. The waveform is a

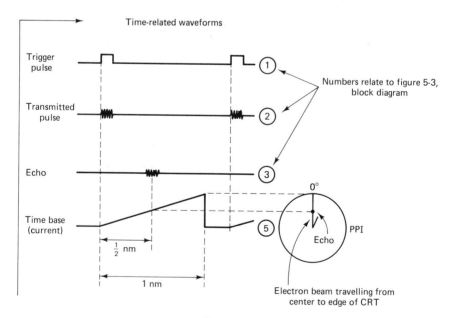

Figure 5–4

linearly rising current waveform (ramp) which will reach the maximum deflection current (the current needed to get the electron beam to edge of the screen) after 12.5 μs which is equivalent, as we already know, to 1 nautical mile. This means if a target is at 1 nm an echo will return after 12.5 μs; it will be amplified by the receiver and then fed as a negative pulse to the cathode of the CRT. This pulse will cause an increase of electrons to be emitted from the cathode and hit the face of the CRT, causing a dot on the screen. The distance away from the center that the dot will occur will be dependent on how far the time-base waveform has deflected the beam. Since this target comes back after 12.5 μs, the timebase has reached maximum deflection current, so the beam will be at the edge of the screen, and that is where the dot will occur.

In the time-related waveform example an echo has returned after 6.25 μs (½ nm). The timebase will therefore have reached ½ maximum deflection current, so the beam will be halfway toward the edge of the screen, and that is where the echo occurs. If the operator is informed that he is on the 1-nm range and that 1 nm exists between the center and the edge of the screen, he can easily calculate by viewing the CRT that the target is at a range of ½ nm.

Please refer now to Fig. 5-5 for the appearance of the three units, the antenna (scanner), the display unit, and the combined transmitter and receiver (transceiver) unit.

Now that we have a good idea of radar principles, two-way travel time, the function and timing of each unit, and how the range and bearing of targets is produced on a CRT screen, we will proceed to introduce and explain satellite communications. A marine radar is explained in greater detail in Chapter 6.

SATELLITE COMMUNICATIONS

The U.S.S.R put the first satellite into orbit in 1957, and this was followed in 1958 by a United States satellite. In 1963 a synchronous-orbit communications satellite (SYNCOM) which had a rotational speed of 24 hours maintained a fixed position with respect to the earth; this means the satellite was in sync or synchronous with the earth's rotational speed. The satellite, which sits approximately 22,300 miles in an orbit above the earth's equator, is equipped with solar cells that provide power for the electronic equipment and retro-rocket systems to occasionally move the satellite so that it remains in its fixed "geostationary position."

To summarize, a satellite is a combination of solar-powered microwave electronics and rocketry, which acts as an unattended relay station.

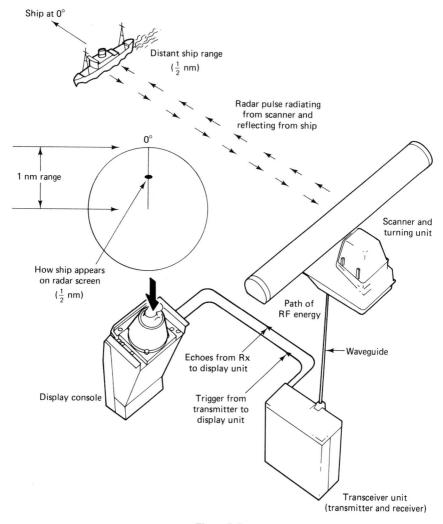

Figure 5-5

USE OF SATELLITES

Satellites are used as relay stations for telephone and digital data transmissions between ships at sea and earth's stations. They are also largely used to link cable television stations to international television networks, and are eventually going to include video shopping for merchandise around the globe, mail transmissions between worldwide post offices, and video conferencing among multiple parties. These are a few of the present and eventual services to be included and provided by satellite communications.

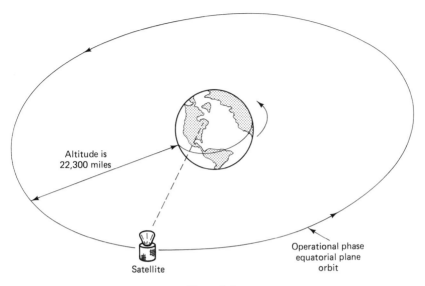

Figure 5-6

The remaining section of this chapter concentrates on introducing you to a marine satellite communications system used for telephone and telex communication between ships at sea and shore-based personnel.

THE MAIN COMPONENTS OF A SATELLITE COMMUNICATION SYSTEM

Refer to Fig. 5-7 for the following explanation of the three devices.

Geostationary Satellite. Orbits above the equator at approximately 36,000 km or 22,300 miles altitude. The satellite serves as a relay transmitter for signals from ship to shore and vice versa. The satellite remains in the same position with respect to the earth, and is stabilized so that the directional antenna is always pointed at the earth. The MARISAT system actually has three satellites in operation, one above the Atlantic Ocean, one above the Pacific Ocean, and one over the Indian Ocean. The coverage areas for these satellites are plotted on a map illustrated in Fig. 5-14.

Shore Station. The shore station connects the shoreside telephone and telex networks to the satellite. It listens continuously on a common request carrier frequency for request bursts from the ship terminals and then controls the terminals and assigns communication channels for each call.

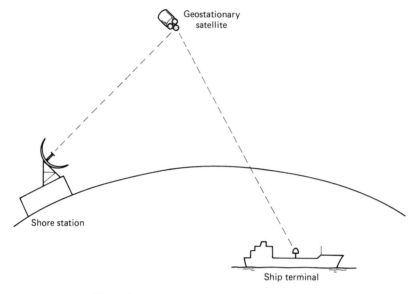

Figure 5-7   Three main components in a satellite.

Ship Terminal.   The terminal connects the teleprinter and the telephones on board the ship to the satellite. The connection is established remotely by the shore station with an assignment message when a call addressed to the terminal is requested from the shoreside networks or when a request burst from the ship's terminal has been acknowledged by the shore station.

SHIP-TO-SHORE COMMUNICATION PROCEDURE

Let's now follow a step-by-step procedure for establishing a telephone link from the ship to the called party via the satellite, earth station, and terrestrial network and establishing a telex link from the teleprinter on the ship to a teleprinter on shore, again via the satellite, earth station, and terrestrial network.

Telephone

The ship's terminal sends a request burst specifying the type of communication to follow (telephone or telex), and waits for the shore station to assign the required communication channels (see Fig. 5-8).

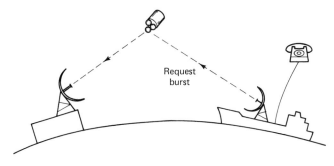

Figure 5-8

   The request burst is transmitted on a carrier common to all terminals and is released by dialing three digits on the ship's terminal.
   The shore station sends the assignment message addressed to the ship's terminal, via the assignment channel of the TDM carrier. (Time Division Multiplexing is explained in Chapter 7.) The assignment message specifies the two voice frequencies (carrier pair) for ship-to-shore and shore-to-ship to be used and tunes the ship's terminal to these frequencies (see Fig. 5-9). While the communication is in progress on the voice carrier pair (Fig. 5-10), the terminal continues to listen to the TDM carrier for other possible messages addressed to the ship's terminal.

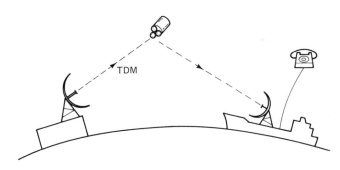

Figure 5-9

Telex

   The ship's terminal sends a request burst specifying the type of communication to follow (in this case telex), and waits for the shore station to assign the required communication channel (see Fig. 5-11). This request burst is released by typing three digits on the teleprinter.
   The shore station sends an assignment message addressed to the ship's

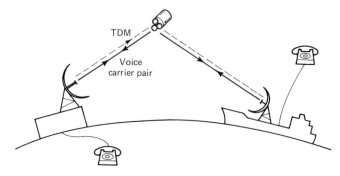

Figure 5-10

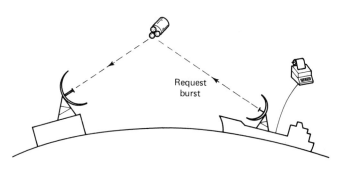

Figure 5-11

terminal via the assignment channel of the TDM carrier. (see Fig. 5-12).
The assignment message specifies the TDM frequency on which the ship
will be receiving telex messages, and the TDMA frequency that the ship will
be transmitting telex messages on. While the communication is in progress
(Fig. 5-13), the ship's terminal continues to listen to the TDM carrier for
other possible messages addressed to the terminal.

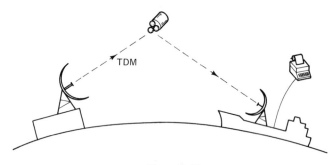

Figure 5-12

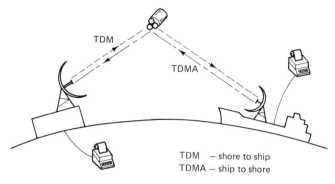

TDM   — shore to ship
TDMA — ship to shore

Figure 5-13

WHY SATELLITE COMMUNICATIONS?

Reliability.   Short wave radio communications is severely affected by variations in ionospheric conditions, different times of the day, and interference. A satellite connection is stable as long as the terminal is within the coverage area of a satellite. Figure 5–14 shows that the three Atlantic, Indian, and Pacific Ocean satellites provide almost worldwide coverage.

Access time.   Radio short wave traffic must pass through a coastal radio station where expedition time may be long under heavy traffic conditions, whereas satellite communications gives almost immediate access to telephone and telegraph networks ashore.

Distress.   This has top priority and allows you to immediately connect to the nearest rescue center regardless of traffic.

Simple operation.   The ship's SAT COM terminal is under the control of the shore station, and therefore requires very little action by the operator.

Privacy.   Short wave communication via a coastal radio station can be picked up by anyone with a short wave receiver; the satellite connection offers full privacy, as an elaborate expensive satellite receiver would be needed to pick up telephone conversations.

We now have covered the principles behind satellite communications; Chapter 7 offers greater depth on the subject.

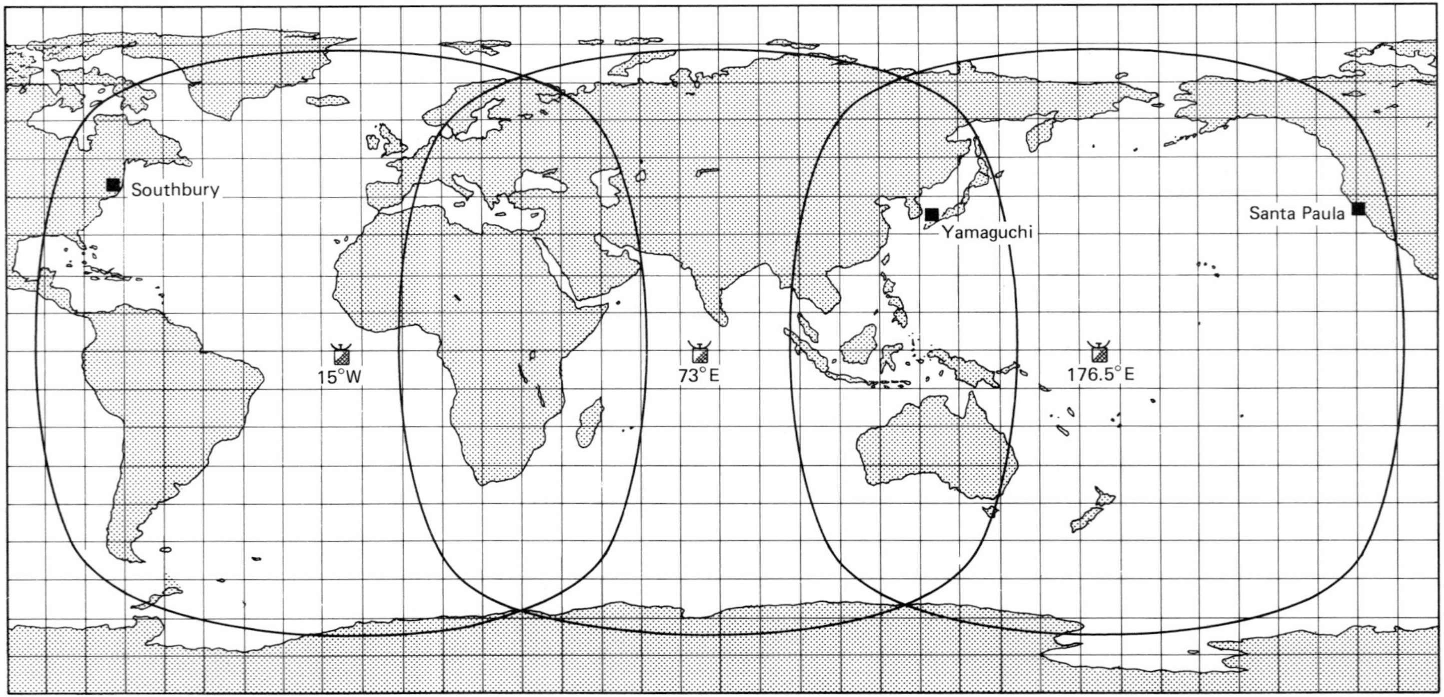

Figure 5-14

SUMMARY

1. This chapter was geared to introducing two microwave systems. We first concentrated on understanding how a microwave pulse of energy takes 12.5 μs to go and return from a target at 1 nm.
2. The basic radar block diagram described how the transmitter produces a trigger pulse and a microwave high-power pulse, the receiver amplifies the echo, and the display unit displays the echo from the target at the correct range and bearing.
3. In the satellite communications section we talked about the three geostationary satellites used for ship-to-shore and vice versa communication.
4. Establishing a link for telephone and for telex communication was almost the same procedure, in that first a request burst was transmitted, then the shore station tuned the ship's terminal to the transmit and receive frequencies, and then communication began.

REVIEW QUESTIONS

1. Sound travels at a speed of:
   (a) 1100 feet per second
   (b) 186,282.397 miles per second
   (c) 162,000 nautical miles per second
   (d) 55 miles per hour
2. The two-way travel time for 1 nm is approximately
   (a) 6.25 μs
   (b) 12.5 μs
   (c) 1 μs
   (d) 25 μs
3. What are the four main blocks in a basic radar block diagram?
   (a) _____
   (b) _____
   (c) _____
   (d) _____

4. Give the full names for the following abbreviations:
   (a) CRT
   (b) PPI
   (c) T/R switch
   (d) Rx
   (e) Tx
   (f) AE
   (g) nm
   (h) SYNCOM
   (i) TDM

5. Calculate the two-way travel times for targets at ranges of:
   (a) 25 nm
   (b) 0.25 nm
   (c) 96 nm

Refer to the basic radar block diagram (Fig. 5–3) for the following questions:

6. The trigger pulse from the transmitter activates two circuits, which then produce their outputs; these are
   (a) _____
   (b) _____

7. The transmitter holds a microwave oscillator which produces a high-power pulsed output. What is the name of this oscillator?

8. What is the purpose of the transmit/receive block?

9. The deflection coils are rotated around the neck of the CRT in synchronism with the antenna's rotational speed. This ensures:
   (a) Correct range of target is displayed on PPI
   (b) Correct bearing of target is displayed on PPI
   (c) Has no importance
   (d) That the antenna is connected to the transmitter for transmission

10. The time-base circuit produces a linearly rising current waveform, which is fed to the deflection coils to:
   (a) Rotate them
   (b) Intensity-modulate the electron beam
   (c) Trigger the transmitter
   (d) Deflect the electron beam from the center of the edge of the PPI

11. Satellites are used as relay stations for:
   (a) Cable TV
   (b) Telephone
   (c) Mail transmissions
   (d) All of the above
   (e) None of the above

12. What are the three main components in a marine satellite communication system?
   (a) _____
   (b) _____
   (c) _____

13. Geostationary satellites:
    (a) Are 22,300 miles above the equator
    (b) Have a rotational speed of 24 hours
    (c) Maintain a fixed position with respect to the earth
    (d) All of the above
    (e) None of the above

14. A telex communication connects:
    (a) A telephone on board ship to a telephone ashore
    (b) A TV signal from ship to shore
    (c) A teleprinter on board ship to a teleprinter ashore
    (d) All of the above
    (e) None of the above

15. Satellite communications has approximately five advantages over short wave (HF) communication. These are
    (a) _____
    (b) _____
    (c) _____
    (d) _____
    (e) _____

RADAR

Objectives

After completing this chapter, you will be able to:

1. Describe the operation and function of a marine radar's scanner, transmitter, receiver, and display unit.

2. Describe the operation of the main blocks and components found inside the four main sections of a radar.

3. Define:
   a. Range rings
   b. Range marker
   c. Synchro bearing transmission
   d. Heading marker

4. Explain all parts of a transmission cycle.

5. Explain duty cycle, minimum range, range resolution.

6. Describe transmit/receive switching operation.

7. Explain what is meant by a multiple trace echo.

8. Define:
   a. Sea clutter
   b. Automatic frequency control

9. Explain time-base generation and bright-up.

10. State the operation and purpose of performance monitors.

11. State the difference between a relative motion and true motion display.

12. Understand the TCMR's display unit, including interference rejection, delay units, and switch mode power supplies.

Radar, as mentioned previously, is an acronym for **radio detection and** ranging, and can be found in the air, on land, and at sea. Airborne radar equipment is utilized for many purposes, examples of which are bombing, navigation, and air search. Harbor radar land-based systems are used to control sea traffic in and out of ports, rivers, and channels. All radar systems operate on the same principles; this chapter is devoted to exploring all aspects concerning marine navigational radars.

When at sea it is imperative that the captain be able to see his or her ship's position relative to land and other vessels in the immediate vicinity.

Radar enables the captain to fix the ship's position by observing reflections from harbors and buoys. It also acts as an anticollision aid which enables operators to calculate the range and bearing of other vessels relative to their own in good or bad visibility.

After our introduction to the principles and main blocks of a radar in Chapter 5, we now proceed to take the transmitter, receiver, and display units and break them up into a more detailed block diagram, which will give us a good idea of what is within each of the main blocks.

Following this is an introduction to the TCMR (typical commercial marine radar), which has been designed to extend our understanding and to help us gain a clear picture of what a typical working radar consists of. This part of the chapter will be split into four sections:

1. Scanner units
2. Transmitters
3. Receivers
4. Display units

Each section discusses how the TCMR scanner, transmitter, receiver and display units achieve the desired results, along with other methods and related information concerning each section.

RADAR BLOCK DIAGRAM

During the next discussion, please refer to both the radar block diagram and the time-related waveforms in Fig. 6–1 and 6–2.

The trigger circuit forms the master timing device for the radar. The trigger circuit produces two positive pulses at a frequency determined by an oscillator within the trigger circuit. This trigger pulse is fed to both the modulator and gate generator. The leading or positive edge of this trigger pulse is a signal to the modulator to now switch through and supply a very high negative DC voltage to the cathode of the magnetron. The modulator is nothing more than a high-voltage switch. The amount of time this switch

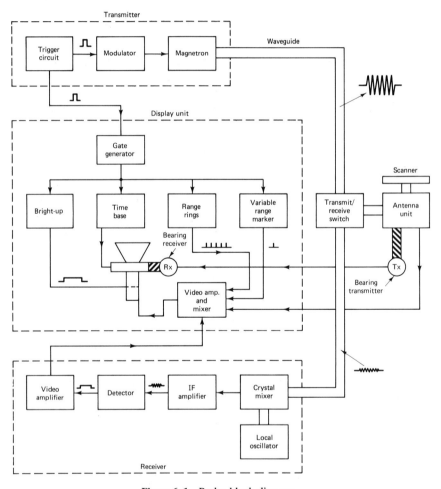

Figure 6–1   Radar block diagram.

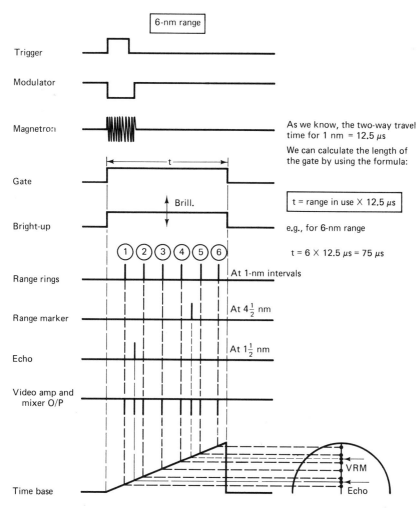

Figure 6-2 Time-related waveforms for radar block diagram.

is closed determines how long the negative DC voltage is applied to the magnetron's cathode and therefore the length of time the magnetron oscillates, as seen in the time-related waveforms.

The magnetron, as we mentioned in Chapter 3, is a high-power, microwave valve oscillator that will oscillate for the duration of the negative high-voltage pulse from the modulator. The microwave pulse from the magnetron is coupled into rectangular waveguide, where it travels to the transmit/receive switch. The energy in the pulse activates the T/R switch, which then routes the pulse to the slotted waveguide antenna, where it is then transmitted into free space.

The antenna rotates at approximately 28 revolutions per minute, which means that it completes 1 revolution (360 degrees) in about 3 seconds. Our oscillator in the trigger circuit can actually produce two or three different frequencies, these frequencies governing the number of pulses per second (pulse repetition frequency—PRF) transmitted from our antenna.

Let us take an example and say that the PRF is equal to 3000 pulses per second. If we transmit 3000 high-powered microwave pulses in one second, and it takes approximately three seconds to complete one antenna revolution, in 360 degrees we would have transmitted 9000 pulses. Dividing 360 degrees into 9000 pps gives us a result of 25 pulses per degree.

Any target in the path of our transmitted pulse will return some of the energy back to our radar scanner. Because of the speed of our pulse to and from the target (speed of light—186,000 miles per second) the sanner has in fact hardly even moved, and the echo is received and fed to the T/R switch.

Any received echoes are of insufficient amplitude to activate the T/R switch and will pass to the mixer in the receiver. The echo frequency is too high to be amplified by conventional circuits, and so a superheterodyne receiver must be used. The local oscillator (L.O.) produces a low-power continuous oscillation which is 60 MHz above the magnetron frequency, and therefore the echo frequency. The L.O. may be a thermionic tube, that is, a Klystron, or a solid-state device such as a Gunn diode. The balanced mixer contains two point-contact or Schottky crystal diodes which will mix the echoes and the local oscillator frequency to produce a sum and difference of the two, both of which will be presented to the intermediate frequency (IF) amplifier. The IF amplifier, which is tuned to the difference frequency, amplifies the echo and passes it to a detector stage. The detector detects and removes the echo information, which is sitting on top of the IF frequency of 60 MHz. This echo information is then amplified by the video amplifier and then fed to the display unit.

The second output from the trigger circuit in the transmitter is sent to the gate generator in the display unit, which produces a waveform as illustrated in the time-related waveforms. The duration of this timing waveform is directly dependent on the range scale presently being used, and is equal to the range in use × 12.5 microseconds.

Example

If the radar operator desires to view a three-mile display, the gate generator's waveform will be

$$
\begin{aligned}
\text{Time} &= \text{range in use} \times 12.5 \text{ microseconds} \\
&= 3 \text{ nm} \times 12.5 \text{ microseconds} \\
&= 37.5 \text{ microseconds}
\end{aligned}
$$

In our time-related waveforms we are using the example and illustrating the waveforms for the 6-nm range. In this situation our gate generators output waveform will last for:

$$
\begin{aligned}
\text{Time} &= \text{range} \times 12.5 \\
&= 6 \times 12.5 \\
&= 75 \text{ microseconds}
\end{aligned}
$$

The gate waveform switches on (when high) and off (when low), the bright-up, time-base, range-ring, and range-marker circuits. The gate generator's output does not control the operation of these circuits, but only the duration for which they are functioning.

The bright-up circuit feeds its output to the grid of the cathode ray tube. This waveform will increase the intensity of the electron beam travelling from the cathode to the face of the CRT when it is positive. A brilliance control on the radar's control panel will adjust the amplitude of this positive waveform and therefore will change the intensity of the electron beam and consequently the brillance of the picture scene on the radar screen.

The time-base circuit produces a sawtooth, linearly rising current waveform that is fed via slip rings to the rotating deflection coils, which are mounted round the neck of the CRT and are rotated in synchronism with the scanner. The electron beam from the cathode travels up the center of the CRT tube and hits the center of the screen, our ship's position. The deflection coils can be moved around the neck of the CRT to any position in the 360 degrees available. When the sawtooth current waveform is passed through these coils, a magnetic field will be set up around the coils and, depending on their position, will determine in which direction the electron beam is deflected from its center position. When the antenna is facing zero degrees, its positional information is fed to the display unit deflection coils, and they are moved to a position so that when the sawtooth current waveform from the time-base circuit is applied, the beam will be deflected in the same direction as the antenna. The rotation of the deflecton coils, and the time-base waveform that is fed to the deflection coils are both separate operations performed by different circuits and are in no way connected to one another. The deflection coils are told to rotate at exactly the same speed as the scanner; if the scanner increases or decreases in speed, the deflection coils are advanced or retarded as necessary. The time-base linear current waveforms are fed at a rate dependent upon the pulse repetition frequency being used, that is, one time-base per transmission, and cause deflection of the spot from the center to the edge of the screen.

As can be seen by the time-related waveforms, bright-up exists only when the time-base is on its positive sweep, that is, from the center to the edge of the screen, and not when the time-base waveform is returning back to its low level. This is to ensure that only the outward sweep of our electron beam is seen and not the flying back of our beam from the edge of the screen to the center of the CRT.

Range rings appear on the display as a set of concentric rings spaced at equal intervals. See Fig. 6–3. As the figure shows, a target appearing on the display without range rings cannot be ranged so easily. With the range rings included on our display, targets can be approximately ranged when they are between two range rings, and accurately ranged when they are on a range ring.

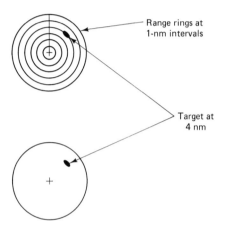

Figure 6–3 Range rings.

For each time-base waveform generated, the range-ring generator produces pulses at fixed intervals of time. In our example, as can be seen in the time-related waveforms, six of these pulses are produced in a 75-microsecond, 6-nm period, so therefore one range ring is generated every 12.5 microseconds, which means that the range rings will occur at a spacing of 1 nm.

If the operator desires an accurate range on a target that lies between two range rings, the range marker can be used. The range marker is one movable range ring that can be set at any radius and placed over any target; the range at which it is set can be read from a display. See Fig. 6–4.

Referring to the time-related waveforms and the block diagram, you will notice that the range marker block produces a delayed pulse which can be positioned at any time along the time-base. In our example, it has been delayed to occur 56.25 microseconds after the start of the time-base, and this pulse will produce a spot on the time-base at a range of 4.5 nm; that is,

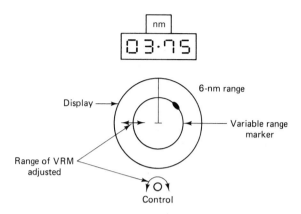

Figure 6-4 The variable range marker (VRM).

$$\frac{56.25}{12.5} = 4.5$$

The range rings and range marker signals are amplified by the video amplifier and mixer and then are fed to the cathode of the CRT as negative-going signals. Every time a negative blip occurs, (be it range rings or range marker), the cathode will be driven more negative, so more electrons will be emitted from the cathode, causing the intensity of the beam to increase. The compositive video intensity modulates the electron beam.

The echoes from the receiver are also fed to the video amplifier and mixer. In our example there is only one echo that returns 18.75 microseconds after transmission; that is, the target is at a range of 1.5 nm. See time-related waveforms.

The output from the video amplifier mixer is known as *composite video* and can be seen in the time-related waveforms.

Let us imagine that our scanner beam is directed towards zero degrees. Therefore our deflection coils have been placed in the necessary position so that when a time-base waveform arrives, the electron beam (the spot in the center of the screen) will be deflected at the same bearing of zero degrees. The time-related waveforms illustrate this time-base waveform, and show how all the video will appear on that time-base as it travels from the screen center to the edge. Let us break down the operation one step further. As you can see, after 12.5 microseconds the time-base waveform has deflected the spot only one-sixth of the way to the screen edge. At this time the first part of our negative composite video, our 1-nm range ring, arrives at the cathode of this CRT and intensity-modulates the beam, producing a paint or blob on our screen.

The range-ring pulses and the range-marker pulse are produced for every time-base through 360 degrees, so they appear on the screen as concentric rings, as seen in Fig. 6–5.

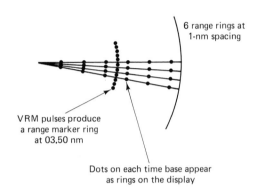

Figure 6–5   Range pulses producing rings.

TCMR UNIT DESCRIPTION AND INTRODUCTION

Figure 6–6 illustrates the TCMR and shows how the separate units are connected, each of which will be discussed in greater detail.

This diagram shows the antenna detecting reflected energy (an echo) from a target at a range of 5 nm and a bearing of 070 degrees. This information is being fed to the display unit, where the ship is appearing on the fifth range ring at a bearing of 070 degrees.

SCANNER UNIT

Included in this unit is a four-foot, center-fed slotted waveguide antenna with a turning motor to rotate the antenna through 360 degrees at 23 revolutions per minute. The horizontally polarized antenna exhibits a vertical beamwidth of 28 degrees and a horizontal beamwidth of less than 2 degrees. The sidelobes are within plus and minus 10 degrees of the main beam, and the antenna's gain with respect to an isotropic radiator is approximately 26 dbs.

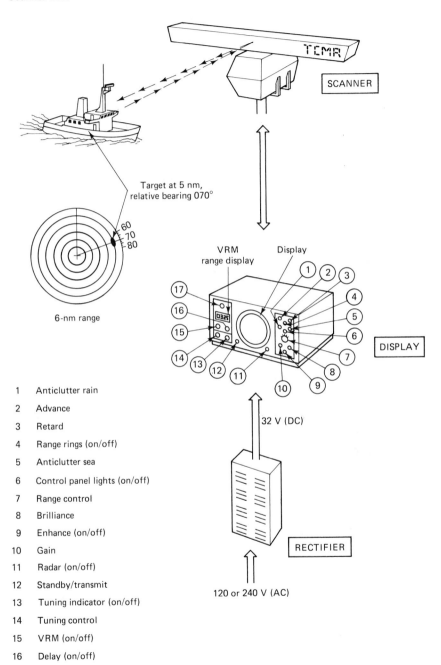

Target at 5 nm,
relative bearing 070°

6-nm range

VRM
range display   Display

SCANNER

DISPLAY

32 V (DC)

RECTIFIER

120 or 240 V (AC)

1   Anticlutter rain
2   Advance
3   Retard
4   Range rings (on/off)
5   Anticlutter sea
6   Control panel lights (on/off)
7   Range control
8   Brilliance
9   Enhance (on/off)
10 Gain
11 Radar (on/off)
12 Standby/transmit
13 Tuning indicator (on/off)
14 Tuning control
15 VRM (on/off)
16 Delay (on/off)
17 VRM range control

Figure 6-6   The TCMR radar system.

A bearing transmitter within this unit sends bearing information to the defelction coils in the display unit, ensuring that they are rotated in synchronism with the antenna.

The transmitter and receiver, known collectively as the *transceiver*, provide the bursts of high-frequency energy transmitted towards targets, and the necessary amplifying circuits needed to process our echoes or reflected energy from surrounding vessels.

The antenna and transceiver, in this case, are incorporated in the same unit and enclosed in a fiberglass casing. Having the transceiver near the antenna ensures a small amount of waveguide run, and so very few losses occur between transceiver and antenna. In the light of the high cost of waveguide, overall costs for the equipment would be greatly reduced.

DISPLAY UNIT

The display unit houses a cathode ray tube and all the circuitry needed to process the received echoes supplied by the receiver and then to present them together with other signals on to the CRT screen. The screen gives all its information in plan view with you in the center of the screen, and everything else relative to your position. The viewed display on the CRT screen is exactly the same as if you were in a helicopter at a high altitude looking at everything around you. This type of display is referred to as a *plan position indicator*. See Fig. 6-7.

The ranger marker/delay unit is part of the display unit and provides

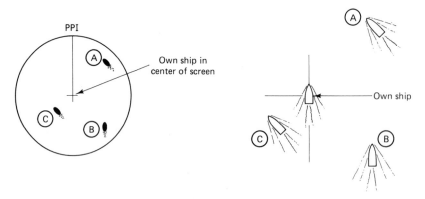

Figure 6-7   (a) PPI display; (b) aerial view.

a very accurate means of measuring the range of a target at any bearing. The range marker is a movable range ring that can be set to any radius and made to cross over a target whose range will now be displayed on the four-digit range display. This unit also incorporates a delay mode which allows the operator to extend the range surveillance capacity of this radar.

RECTIFIER UNIT

This unit converts the ship's AC supply of 120 or 240 volts to 32 volts of direct current to drive the TCMR's power supply and scanner turning motor.

TCMR OPERATING INSTRUCTIONS

The operation, use, and correct setting of all these controls are discussed separately as we come across them when covering the system in greater detail.

TCMR SYSTEM DESCRIPTION

The TCMR block diagram is illustrated in Fig. 6–8, and should be referred to during the explanation of each section.

The striped broken lines encompass and separate the two main units, that is, display and scanner unit. The scanner unit assembly houses a separate transceiver assembly on which is mounted the trigger board outlined by the solid broken lines.

The display unit has the range marker/delay unit and three boards mounted in it. These boards, outlined by the solid broken lines, are called the *power supply control, main display,* and *interference rejection board.*

During our explanation of each block and device we will refer to and gain a good understanding of how the TCMR achieves these desired functions, and then cover in general other methods and new ideas relating to that area of discussion.

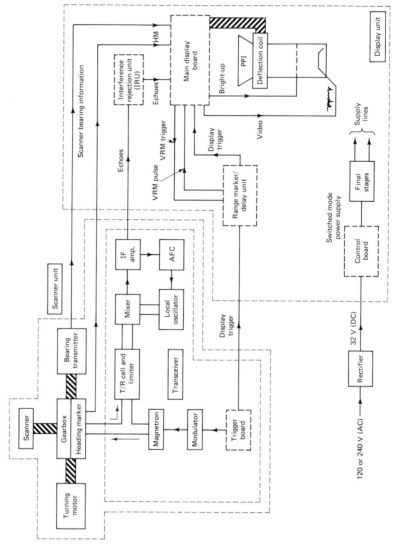

Figure 6-8 TCMR block diagram.

124

REVIEW QUESTIONS FOR INTRODUCTION

Refer to Fig. 6-1.

1. The trigger circuit:
   (a) Is a switch connecting high voltage through to the magnetron
   (b) Is the master timing device of the radar
   (c) Is a microwave frequency oscillator
   (d) Receives bearing information from the scanner

2. What is the function of the T/R switch?

3. The gate generator drives what four circuits?
   (a) _____
   (b) _____
   (c) _____
   (d) _____

4. In which block would a Gunn diode be found?

5. The _____ antenna completes one revolution in approximately _____ seconds.
   (a) Parabolic, 25
   (b) SWG, 15
   (c) SWG, 3
   (d) Parabolic, 20

6. What is the function of the time-base circuit?

7. The IF amplifier is tuned to the _____ the local oscillator and echo frequency.
   (a) Sum of
   (b) Difference between
   (c) Both (a) and (b)
   (d) None of the above

8. What is the purpose of range rings on the CRT screen?

9. If the operator has selected a 48-nm range, and six range rings are on the screen, their spacing would be one every:
   (a) 1 nm
   (b) 6 nm
   (c) 8 nm
   (d) 14 nm

10. What advantages does the variable range marker have over the range rings?

11. The TCMR radar system (Fig. 6-6) has three main units, which are
    (a) _____
    (b) _____
    (c) _____

12. What is the function of the rectifier unit?

13. The scanner unit houses the
   (a) Scanner, transmitter, and display
   (b) Transmitter, receiver, and CRT
   (c) Scanner and transceiver
   (d) None of the above
14. What is meant by a PPI display?
15. PPI is an abbreviation for_____.

SECTION 1 SCANNER UNITS

TCMR Antenna and Turning Unit

A turning motor is supplied with 23 volts DC from the power supply in the display unit, and its function is to turn the 4-foot SWG antenna at 23 revolutions per minute.

A bearing transmitter, as can be seen in the TCMR block diagram, is mechanically linked to the antenna to monitor the scanner's rotational speed and feed this speed information to the rotating deflection coils in the display unit. A simplified diagram of the bearing transmitter system is shown in Fig. 6-9.

A four-pole magnet is mechanically coupled to the scanner via a gearbox. As the scanner is rotated, so is the four-pole magnet, and when each pole passes by the inductor, pulses will be induced into the coil and then applied to the base of the amplifying transistor. The output on the collector of the transistor is a series of pulses in step with the scanner speed, the frequency of which is a measure of the scanner's rotational rate. These

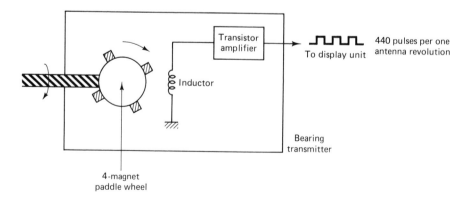

Figure 6-9   TCMR bearing transmitter.

pulses are sent down to the deflection coil motor in the display unit, and will synchronize the rotating deflection coils to the rotating scanner.

The four-pole magnet will actually rotate 110 times in one complete scanner rotation. For every one revolution of this magnet, four pulses are induced and fed down to the display unit. Consequently, a total of 440 pulses will be received by the display unit for a complete scanner rotation through 360 degrees.

An alternative to the TCMR's bearing transmission system is the synchro or servo system, a description of which now follows.

Synchro or Servo Bearing Transmission Systems

A synchro is a motorlike device that is capable of transforming mechanical angular position into an electrical output which represents the angular displacement. When two or more synchros are connected together to a source of AC voltage, they form a synchro system which can be utilized to rotate the deflection coils in the display unit in synchronism with the scanner.

A synchro system transmits mechanical information without amplification; that is, power out is equal to power in. A servo system provides amplification of the mechanical information so that a larger load can be driven; it uses synchros.

A Simple Synchro System

A synchro device contains a stator and rotor, the latter of which is free to rotate.

As can be seen in Fig. 6-10, an energizing AC voltage is applied to the rotors of the synchro transmitter (TX) and the rotor of the synchro receiver (TR).

The scanner, via a gearbox, is mechanically connected to the rotor of the synchro transmitter. The stator of the synchro transmitter is electrically connected to the stator of the synchro receiver. The rotor of the synchro receiver is mechanically coupled to the free-to-move deflection coils.

Consider the rotors in exactly the same position relative to their stators. The AC energizing voltage in the TX rotor induces voltages into the TX stator windings, and as the rotors are in exactly the same position, these same voltages will also be induced in the TR stator windings by the TR rotor. No current will flow between the two stator windings, as no potential difference is present, and a state of equilibrium or balance will exist with the deflection coils aligned to exactly the same position as the scanner.

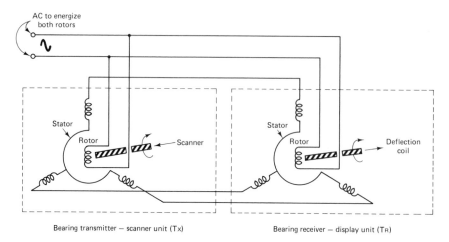

Bearing transmitter — scanner unit (Tx)   Bearing receiver — display unit (TR)

Figure 6-10   A synchro bearing transmission system.

If the TX rotor is mechanically turned through a few degrees, the voltages in the TX stator windings will no longer equal the voltages in the TR stator windings. A potential difference now exists, so current flows between the two stators, producing magnetic fields in the TR stator windings. The resultant stator magnetic field in the TR will interact with the rotor's AC energizing field, producing a turning force or torque. The TR rotor will turn until the torque is removed, which occurs when the rotor has taken up the same relative position as that of the TX rotor; that is, the deflection coils are aligned to the same position as the scanner.

Heading Marker

The heading marker indicates the direction of the fore and aft line of the ship, that is, the ship's heading, by placing a straight line on the PPI originating from the center of the screen at a bearing of 0 deg, as seen in Fig. 6-11.

Referring back to our TCMR system block diagram, you will notice that the heading marker block is mechanically connected to the scanner. The heading marker signal is generated by a magnet mounted on the rotating scanner. When the scanner is pointing to 0 degrees (straight ahead), the magnet is alongside a reed switch mounted permanently in the aerial turning unit. The magnet passing by the switch causes the contacts within the reed switch to close. One side of the reed switch is connected to the chassis or 0 volts; the other is connected to a heading marker circuit in the

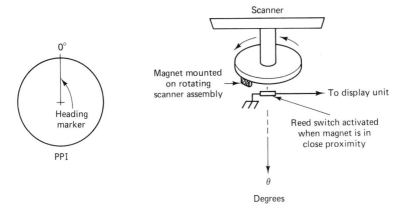

Figure 6–11 The heading marker.

display unit which will generate the necessary signal to produce the heading line. As the aerial sweeps through the direction of the vessel's heading, the switch closes and 0 volts is fed down to the heading marker circuit, triggering it into operation. Normally the line floats at something other than 0 volts, but when triggered it is taken to 0 volts.

REVIEW QUESTIONS FOR SECTION 1

1. The TCMR bearing transmitter will transmit _____ pulses to the display unit for one revolution of the antenna.
 (a) 840
 (b) 750
 (c) 440
 (d) 375

2. What is the difference between a synchro and a servo system?

3. Give the full names for the following abbreviations:
 (a) TX
 (b) TR
 (c) Tx
 (d) Rx

4. What is the name of the voltage applied to the rotors of a synchro transmitter and receiver?

5. The bearing transmitter sends the scanner position information to the
   _____ in the _____ .
   (a) Gate generator, display unit
   (b) Time-base, display unit
   (c) Deflection coil, display unit
   (d) Grid of the CRT, display unit

6. What is the function of the heading marker line on the PPI?

7. How many heading marker lines are produced in one revolution of the scanner?
   (a) 10
   (b) 1
   (c) 3
   (d) 4

8. Referring to Fig. 6–11, explain the function of the reed switch?

9. The deflection coils must be rotated in synchronism with the antenna to ensure correct
   (a) Range of targets displayed
   (b) Amplitude of echoes
   (c) Frequency of transmission
   (d) Bearing of targets displayed

10. The system that ensures that the deflection coil is rotated in synchronism with the antenna is called
    (a) The heading marker system
    (b) The range marker unit
    (c) The bearing transmission system
    (d) The interference rejection unit

SECTION 2 TRANSMITTERS

Before discussing the TCMR transmitter, we must first become familiar with a few abbreviations, terms, and principles associated with all radar transmitters.

Transmission Cycle

Figure 6–12 illustrates the transmission cycle for a typical radar set.
As we already know, we are dealing with pulsed systems and therefore it is importanat that we understand certain terms associated with pulse radar transmitters.

Peak power (PP.)   The maximum power of the pulse from a radar transmitter.

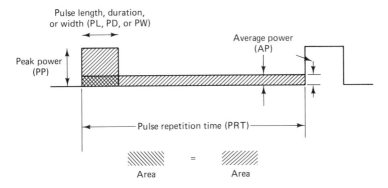

Figure 6-12 Radar transmission cycle.

Pulse duration. Also called *pulse length* (PL) or *width*, this is the time between the leading and trailing edges of the peak pulse amplitude.

Average power. This is the power delivered to the transmitter averaged out over a complete cycle.

Pulse repetition frequency (PRF) or **Pulse Repetition Rate (PRR).** This is the rate at which pulses are transmitted from a radar set, and is usually given in hertz or pps (pulses per second.)

Pulse repetition time. Abbreviated PRT, and also known as PRP (pulse repetition period), it is the reciprocal of PRF and is a measure of the time between pulses.

The maximum range of our radar is proportional to the average power. With this in mind, let us go ahead and analyze all the factors relating to average power.

The area within the pulse is equal to PP × PL, and if this area were spread out over the complete repetition time, we would gain a value of average power for the transmission cycle; that is,

$$\text{Peak power} \times \text{pulse length} = \text{prt} \times \text{average power}$$

The area within the pulse is equal to the average power area over the complete cycle.

By transposition, we can acquire a formula for average power:

$$\text{Average power} = \frac{\text{peak power} \times \text{pulse length}}{\text{prt}}$$

If average power is inversely proportional to pulse repetition time, it will be proportional to pulse repetition frequency, so a final formula can be acquired:

Average power = peak power × pulse length × prf (watts)

Example

Peak power = 3 kw   prf = 850 pps   pulse length = 0.75 μs
    AP = PP × PL × PRF
       = 3 kw × 0.75 μs × 850
       = 1.91 watts

Duty Cycle

This is the measure of time for which the transmitter is transmitting or working, and is equal to the pulse length × the pulse repetition frequency.

The higher the duty cycle, the greater the heat developed in the transmitter, and the greater the power rating of the components.

Example

Pulse length = 1 μs   PRF = 500 pps
Duty cycle = pulse length × PRF
       = 1 μs × 500
       = 0.0005

If we multiply by 100, we find that the transmitter is working for 0.05 percent of the time, so in a period of one hour (3600 seconds) the transmitter would have been working for

0.05% of 3600 seconds

$$\frac{0.05}{100} \times \frac{3600}{1} = 1.8 \text{ seconds}$$

Minimum Range

The trigger from the trigger board activates two circuits:

1. The transmitter, which is told to transmit its high-power microwave frequency pulse.
2. The display unit, which is told to begin a time-base sweep from the center to the edge of the screen.

All the time we are transmitting, we cannot receive echoes, and the minimum range formula allows us to calculate the closest range at which we will receive targets.

Minimum range is determined by the time it takes from the commencement of the transmitter pulse for the receiver to become capable of receiving echoes. The receiver is rendered inoperative during transmission, and the transmit/receive switch normally needs a small amount of recovery time after transmission before it has switched and now is allowing echoes to travel from the antenna to the receiver.

Min. range = (pulse length μs + T/R recovery time μs) × 150

Example (Fig. 6–13)

$$PL = 0.05 \ \mu s \qquad T/R \text{ recovery time} = 0.1 \ \mu s$$
$$\text{Min. range (meters)} = (PL \ \mu s + T/R \text{ recovery } \mu s) \times 150$$
$$= (0.05 \ \mu s + + 0.1 \ \mu s) \times 150$$
$$= 0.15 \times 150 = 22.5 \text{ meters}$$

For a good minimum range value, the pulse length must be as small as possible and the T/R switch must be fast-acting. The vertical beamwidth of the antenna must be sufficient to prevent close targets from passing under the beam.

Range Resolution

Range resolution is the ability of the radar to discriminate between two small objects close together on the same bearing, that is, to produce separate echoes.

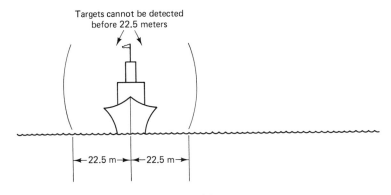

Figure 6–13 Minimum range.

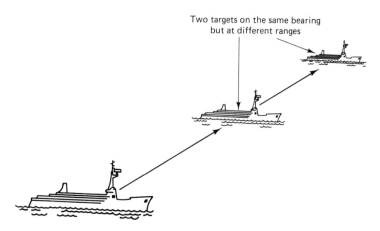

Figure 6-14 Range discrimination.

Range resolution is completely dependent on the pulse length, and is completely different from bearing resolution, which was discussed in relation to antennas. See Fig. 6–14.

With reference to Fig. 6–15:

(a) You will notice that we are transmitting a pulse of 0.2 μs, which will travel at the speed of light, that is, 300,000,000 meters per second, and will therefore occupy 60 meters as it travels toward two beacons positioned 30 meters apart in range.

(b) The 60-meter pulse has arrived at the first beacon and it begins to return or reflect some of the energy back to the ship.

(c) Here we notice our pulse striking our second beacon and causing it to begin reflecting back energy.

(d) Our pulse is now leaving our first beacon, and no more energy is reflected from our beacon.

(e) Similarly, our pulse is now leaving our second beacon, and our 120-meter continuous return is now travelling back to our ship, as illustrated in Fig. 6–15 (f).

The display unit screen in Fig. 6–15 (g) illustrates the solid one return from our two beacons.

By shortening the transmission pulse to 0.05 μs (15 meters), we will receive two separate reflections from each beacon, as our pulse now fits between the two targets. See Fig. 6–15 (h).

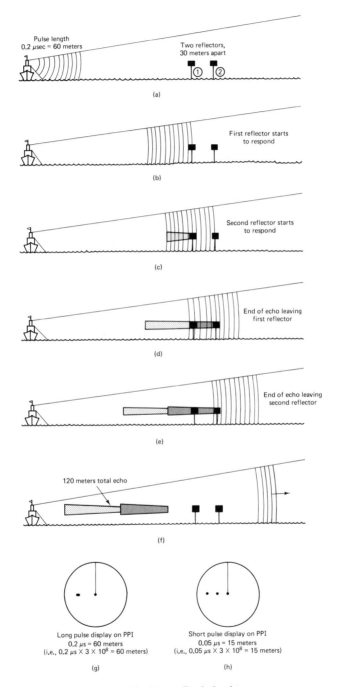

Figure 6–15 Range discrimination.

T/R Switching (Transmit/Receive)

Most radars at sea employ a common antenna and microwave frequency feeder to the antenna for the transmitter and the receiver; that is, both transmitter and receiver use a common antenna.

The receiver operates with a power input measured in microwatts, while the transmitter radiates an output in kilowatts. A switch must be utilized that is fast enough to protect the receiver as the transmitter turns on, and fast enough to allow received signals to the receiver the moment the transmitter pulse ends.

Figure 6–16 shows a T/R switch connected in a radar system. A spark gap in a gas-filled cavity is used as the T/R switch. The cavity is tuned to the transmitted frequency by adjusting the gap between electrodes. Pre-ionization from a "keep alive" electrode at the center of one of the electrodes ensures a very fast switch-on time, while the gas that gives a minimum recovery time is chosen.

When the transmitter fires, the gap between the electrodes breaks down, causing a spark which represents a short circuit across the gap terminals. As can be seen in Fig. 6–16, the T/R switch is so positioned so that a short circuit across its electrodes will present an open circuit at the junc-

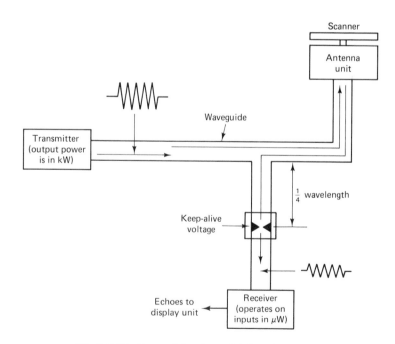

Figure 6–16   Transmit/receive switch in radar system.

tion point a quarter of a wavelength away. The large power pulse from the transmitter's magnetron will therefore view the path to the receiver as a very high impedance, so all of the energy will travel directly to the antenna.

After the transmitter pulse, the spark gap becomes an open circuit and so reflects a short circuit across the junction points. The received power, upon reaching the junction, will therefore follow the low-impedance path, presented by the T/R switch, to the receiver.

Selection of Pulse Repetition Frequency

After discussing the previous subjects, which are concerned with transmitters, we can now analyze some of the factors which govern what PRF is chosen for what range scale:

1. The PRF must be such that sufficient time be allowed between successive transmissions for an echo to return from a target at maximum range. That is, the PRF must be small so that the PRT (time between transmissions) is large for long ranges, so that targets at maximum range have enough time to return before another transmission occurs.
2. PRF is one of the factors determining the maximum peak power, and therefore the power-handling capabilities must be taken into consideration when one is deciding which PRF to use, that is, which duty cycle.
3. The higher the PRF, the higher the average power, and therefore the greater the range of the radar. It is the average power that determines the range performance of a radar.
4. When an antenna is rotated at a constant speed, the PRF must be such that a sufficient number of pulses strike the target during each antenna revolution to ensure adequate echo returns (minimum six). That is, let's assume that the scanner rotation rate and PRF are such that six transmissions and therefore six echoes are returned from a target. If the PRF is halved, only three returns will result from that same target, and this will be insufficient to produce a visible paint on the CRT screen. The number of pulses striking a target therefore depends on the horizontal beamwidth, the PRF, and the scanner rotation speed.
5. From the previous example in 4, it is easy to imagine that a high PRF would produce a bright picture.

Summary. Because of the reason given in item 1, a low PRF is chosen on long ranges to ensure a long PRT so that echoes from targets at long ranges will have sufficient time to return before another transmission occurs; on short ranges a high PRF is chosen.

Selection of Pulse Length

1. The shorter the pulse length transmitted, the better the range resolution acquired.
2. A shorter transmitted pulse length will give the best minimum range.
3. Pulse length in conjunction with PRF determines the maximum peak power; that is, the duty cycle and pulse length must be selected so as not to exceed duty cycle figures for components.
4. The maximum range of a radar depends on the average power radiated; thus for maximum range, the pulse should be as long as possible.

Summary.   Short pulses are used on short ranges for good minimum range and range resolution, and long pulses on long ranges are used to increase the average power of our radar system, and hence the range.

Multiple-Trace Echoes

Let us first imagine, as an example, a radar system operating with a PRF of 2000 pulses per second, and show the information relating to that radar's transmission cycle, as seen in Fig. 6–17.

As can be seen in Fig. 6–17(a), with a PRF equal to 2000 pps, our interpulse period (PRT) calculates to be 500 μs. We have selected the 30-nm range, which means that the time-base will last for approximately 375 μs, after which a delay of 125 μs occurs. This period is referred to as *dead time,* as nothing is active.

Our first transmission sends out a pulse of microwave energy as usual, which travels off at a bearing of 270 degrees, as seen in Fig. 6–17(b). Two targets, *A* and *B,* are present at 10 and 20 nautical miles, respectively, so echoes return to our radar and are displayed by our time-base at their correct ranges of 10 and 20 nm at a bearing of 270 degrees by the time-base. No targets are present between 30 and 40 nm, so no echoes are returned; even if there were echoes returned, they would not be displayed, because the time-base has now sent the electron beam to the center of the screen during dead time. However, there is a target at 50 nm which will cause some of the original first transmission's power to return back to our radar. As illustrated in the time-related waveforms (Fig. 6–15), it will actually return

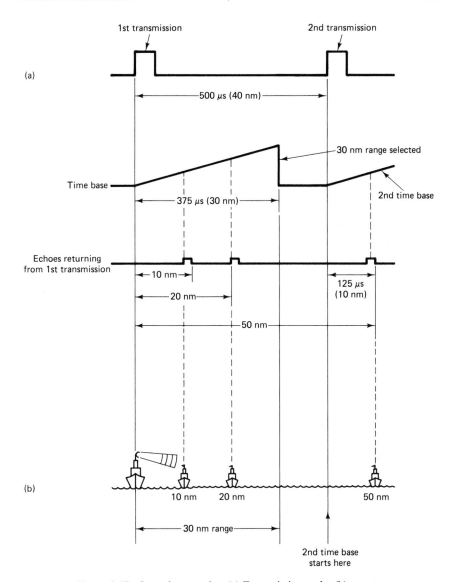

Figure 6-17 Second-trace echo. (a) Transmission cycle; (b) targets.

125 μs after the second transmission. At this time the second time-base has already reached a range of 10 nm of the CRT screen, so our 50-nm target will be presented on the screen at a range of 10 nm.

It would be desirable to be able to distinguish between a correct return, and a multiple-trace return. Some radars, such as the TCMR, utilize

a wobbulator (sweep generator) which causes a small change in the PRF
frequency.

Example

Figure 6–18 shows how the PRF has been decreased by the wobbulator to 1777.7
pulses per second. The PRT, which is the reciprocal of PRF, calculates to be 562.5
μs. Our 10- and 20-nm targets will not be affected by this change, but our 50-nm
target, which returns 625 μs after transmission, will now occur on the second time
base after a time of

$$625 \ \mu s \ - \ 562.2 \ \mu s \ = \ 62.5 \ \mu s$$

This means our 50-nm target will be displayed on the second time base at a range
of

$$\frac{62.5 \ \mu s}{12.5 \ \mu s} = 5 \ nm$$

Our 50-nm target was originally displayed, when the PRF was 2000, at a range
of 10 nm. Now our PRF has been shifted down to 1777.7; the 50-nm target has
moved to 5 nm, but our true targets that occur within the range we selected are still
at 10 and 20 nm.

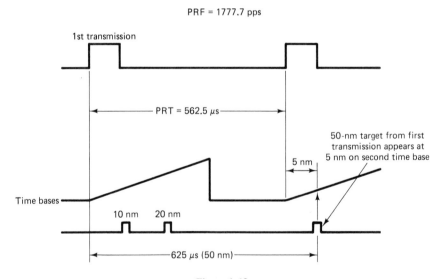

Figure 6–18

Figure 6–19 illustrates how the wobbulator has now swept the PRF up to
2285.7 pps, which means that our PRT will change once again but will now equal
437.5 μs. This change again will have no effect on the 10- and 20-nm targets, but

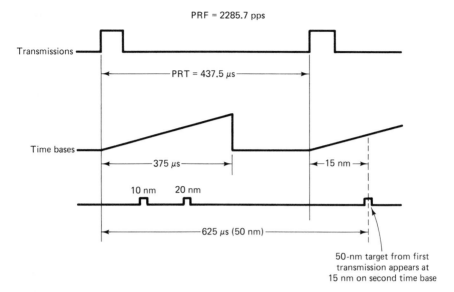

Figure 6-19

the 50-nm target which returns 625 μs after transmission will appear on the second time base after

$$625 \ \mu\text{s} \ - \ 437.5 \ \mu\text{s} \ = \ 187.5 \ \mu\text{s}$$

This time will equal in range:

$$\frac{187.5 \ \mu\text{s}}{12.5 \ \mu\text{s}} \ = \ 15 \ \text{nm}$$

     This example has shown that when we vary the PRF, our first-trace true echoes are not affected, but second- or multiple-trace echoes are shifted back and forth in range, which gives them a woolly appearance on the PPI, as shown in Fig. 6-20.

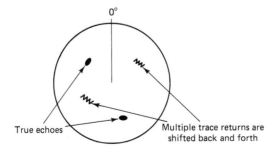

Figure 6-20   True and false (multiple-trace) echoes.

TCMR Transmitter

Referring to the TCMR system block diagram (Fig. 6–8) once again, you will notice that the transceiver chassis contains all the electronic circuitry within the scanner unit with the exception of the heading marker and bearing transmitter.

Figure 6–21 illustrates a simplified block diagram of the TCMR's transmitter.

The PRF generator, which generates the triggers, is a free-running oscillator which immediately begins oscillating the moment the radar is switched on. It produces triggers at two PRF rates of 1500 or 750 pulses per second. The PRF desired would be selected by the pulse-length control input to the transceiver board, which originates, and is determined by, the range selected on the display unit; that is,

Range Selected	PRF	Pulse Length	PRT
0.25, 0.75, 1.5.	1500 Hz	0.08 μs (S/P)	666 μs
3, 6, 12, 24, 48	750 Hz	0.55 μs (L/P)	1333 μs

There are three outputs from the PRF generator; two of these are fed to the modulator part of the trigger board, to produce a charge and a discharge trigger.

The charge trigger is fed from the PRF generator to the charge driver via a gate. The charge driver produces a negative output trigger that is applied to the charge circuit, which switches through a large voltage of 130 volts to charge the pulse-forming network.

Half a PRF cycle later, triggers occur on the other two outputs of the PRF generator. The second output drives the SCR (silicon control rectifier) transmit switch, which produces an output pulse that is applied to the pulse-former as a discharge trigger. This discharge trigger fed to the pulse-forming network initiates the negative output pulse to the magnetron, the length of this output pulse being controlled by the pulse length (PL) input. The magnetron is a high-power (10 kw) microwave oscillator (operating in the region of 9340 MHz) and will oscillate for the duration of the negative voltage pulse applied to its cathode. The burst of microwave energy from a magnetron is then coupled into the waveguide. This pulse activates the T/R cell, which then routes the pulse to the scanner.

The third and final output from the PRF generator is fed down to trigger the display unit via the gate and display trigger circuit. This final trigger out can be derived from two paths:

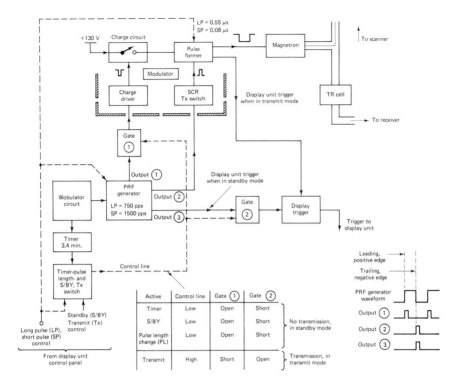

Figure 6–21 TCMR transmitter block diagram.

1. When the TCMR is set in its standby mode, everything is operational but no transmission takes place, and the CRT screen displays everything except echoes. This condition is also known as *immediate readiness*, as the radar set is fully operational, but we are not transmitting and wasting the relatively short life expectancy of most magnetrons. This mode is often chosen when a display of the surrounding area is not desired at present, but could be needed at any time. Referring back to the block diagram in Fig. 6–21, you will notice that a long pulse/short pulse input and standby/transmit input are fed from the display unit controls to a block entitled "Timer, PL, standby/transmit." A control line from this block is fed to gate ②, and enables the trigger through from the PRF generator only when any one of the three conditions is met:

   a. Standby is selected on the control panel.

   b. If a pulse-length change occurs when changing ranges. The radar is automatically put in the standby mode until the transmitter has settled to transmitting its new pulse length.

c. In the first three and a half minutes after switching on the radar set, a timer comes into operation and ensures that no transmission occurs within this time frame. During this period of time the equipment is held in standby mode to allow the magnetron to warm up before the large negative voltage is applied to the cathode.

2. When transmit is selected on the control panel of the display unit, the gate allowing the trigger from the PRF generator to the display trigger circuit is disabled and now the display unit trigger is derived from a tapping in the modulator pulse-forming network assembly. The display unit trigger is derived from this point rather than directly from the PRF generator due to a timing difference occurring between the PRF generator's output to start transmission, and the actual transmission's occurring, thus ensuring synchronization of the display unit to the firing of the transmitter.

In the standby mode transmission must be stopped. This is achieved by the same control line inhibiting a gate which breaks the charge trigger to the charge driver. Although a discharge trigger is still produced, the pulse-forming network has not been allowed to charge, and therefore no transmission will occur.

A wobbulator of 40 Hz has two outputs:

1. One is fed to the timer, which at switch-on is reset, and begins to count the 40-Hz pulses. After a period of about 3.4 minutes, the timer produces an enable signal to allow transmission to occur if we have selected to transmit. This 3.4-minute delay at switch-on is to ensure that the magnetron has enough time to warm up before the very large negative DC pulse is applied to its cathode. Without this warm-up period, excessive damage to the magnetron and a short life span results.

2. The 40 Hz is also applied to the PRF generator to vary its frequency and enable the operator to identify second-trace echoes.

REVIEW QUESTIONS FOR SECTION 2

1. Give the full names for the following abbreviations:
   (a) PL
   (b) PD
   (c) PRF

    (d) PRT

    (e) AP

    (f) PW

    (g) PP

2. Calculate the average power when:

    PP = 70 kw   PL = 0.12 μs   PRT = 600 μs

3. The T/R switch:

    (a) Allows us to employ a common antenna for transmission and reception

    (b) Protects the receiver during transmission

    (c) Allows received echoes to pass to the receiver

    (d) All of the above

    (e) None of the above

4. Explain the main reason why PRF is low on long ranges.

5. A short pulse is chosen on short ranges.

    (a) True

    (b) False

6. A long pulse is chosen on long ranges to increase average power, and therefore range.

    (a) True

    (b) False

7. A short pulse transmission on short ranges

    (a) Provides good minimum range

    (b) Provides good bearing resolution

    (c) Provides good range resolution

    (d) Both (a) and (b)

    (e) Both (a) and (c)

8. Explain briefly what is meant by a multiple-trace return.

9. A wobbulator circuit is used to:

    (a) Vary the radar's pulse length

    (b) Vary the radar's PRT

    (c) Increase peak power

    (d) None of the above

10. Refer to Fig. 6–20. The function of the timer is to ensure:

    (a) No transmission for the first 3.5 minutes.

    (b) The magnetron has a warm-up period to prevent damage

    (c) The total number of transmissions does not exceed 22

    (d) Both (a) and (c)

    (e) Both (a) and (b)

11. If the 1.5-nm range is selected by the operator, the long-pulse/short-pulse control line from the display unit will select a PRT of _____ and a pulse length of _____ .

    (a) 1500 Hz, 0.55 μs

    (b) 750 Hz, 0.08 μs

    (c) 1500 Hz, 0.08 μs

    (d) 750 Hz, 0.55 μs

12. If transmit is selected by the operator, and no transmission occurs:
    (a) We are in the first 3.5 minutes after switch-on
    (b) A pulse length change has just occurred
    (c) Both (a) and (b)
    (d) None of the above

13. Transmission is stopped in the standby mode by:
    (a) Inhibiting the transmit trigger
    (b) Inhibiting the charge trigger
    (c) Opening up gate 2
    (d) Changing the PRF

14. During standby, the display unit trigger is derived from:
    (a) A pulse former
    (b) The magnetron
    (c) Output 3 of the PRF generator
    (d) All of the above

15. If the PRF = 1500 Hz, then the PRT = _____ .
    (a) 666 μs
    (b) 1333 μs
    (c) 300 μs
    (d) 1200 μs

SECTION 3 RECEIVERS

TCMR Receiver

Refer to Fig. 6–22 for the following explanation of the TCMR receiver block diagram.

The radio frequency echo pulses reflected by a distant object are similar to the transmitted pulses in frequency and shape, but are considerably diminished in amplitude. These minute signals are amplified and converted into video pulses by the receiver. A voltage amplification in the order of 10,000,000 is required to produce a video pulse of sufficient amplitude to intensify the beam of the CRT. The TCMR receiver must accomplish this amplification with minimum introduction of noise voltages.

In addition to having a high gain and a low noise figure, the TCMR receiver must provide a sufficient bandwidth to pass the many harmonics contained in the video pulses. That is, the bandwidth is increased for short-pulse transmission, as more harmonics are present, so a larger bandwidth is needed to preserve the shape of the pulse, and therefore to minimize distortion of the pulses. The receiver must also accurately track, or follow, the transmitter in frequency, since the amount of drift will diminish the reception of the echo signal.

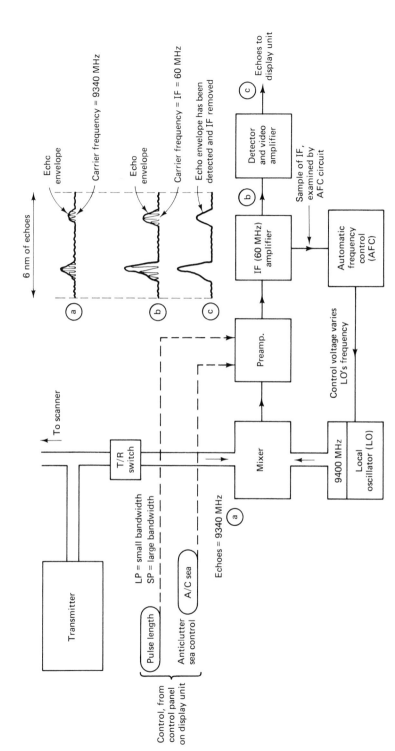

Figure 6-22 The TCMR receiver block diagram.

In the TCMR block diagram, the transmitter is sending high-power pulses at a microwave frequency of 9340 MHz. These high-power pulses cause the T/R cell to arc, which then routes our pulses directly to the antenna and then out into free space. Received low-power echoes from targets are of insufficient amplitude to activate the T/R cell and pass directly down to the balanced mixer. Here, the echoes are mixed with a local oscillator frequency of 9400 MHz, resulting in a sum and difference at the output of the mixer.

Echo Frequency	Local Oscillator Frequency
9340	9400

$$\text{Sum} = 9340 + 9400 = 18{,}740 \text{ MHz}$$
$$\text{Difference} = 9400 - 9340 = 60 \text{ MHz}$$

The whole purpose of the mixer in a radar system is to lower the echo frequency so that it can be amplified by conventional circuits. Our desired frequency is therefore the difference or intermediate frequency of 60 MHz, which will be selected by tuning the first stage of the IF amplifier to this frequency.

The IF amplifier achieves our very high gain needed to convert our microvolt input echo pulses to usable video, which is then fed down to the display unit.

The local oscillator and the magnetron both have resonant cavities to cause and determine their oscillation frequency. During the first hour of operation, the radar goes from a cold to a warmed-up condition; because of this, the cavities expand as time elapses. If any change in frequency occurs in either the magnetron or local oscillator, the difference will no longer be equal to 60 MHz. As the IF amplifier is tuned only to this frequency, another frequency will not be amplified and passed to the display unit. The automatic frequency control (AFC) circuit takes a sample of the intermediate frequency and analyzes it to ensure that it is exactly at 60 MHz. The AFC circuit outputs a control voltage which is used to vary the frequency of the local oscillator. If, for example, the magnetron were to increase frequency to 9350 MHz, the IF would now be the difference between 9350 MHz and 9400 MHz, which is a downward sweep from 60 to 50 MHz. This sweep would be sensed by the AFC circuit, and the control voltage would be changed to increase the frequency of the local oscillator to 9410 MHz, so the difference would still be equal to 60 MHz and therefore would be amplified by the IF amplifier. This is a closed-loop system which continually senses and automatically corrects, to ensure that the IF is locked at 60 MHz. A further explanation of AFC with its advantages is discussed at a later time in this section.

IF Amplifiers in General

The IF section of a microwave receiver determines the receiver's gain, signal-to-noise ratio, and effective bandwidth. The IF amplifier stages must have sufficient gain and dynamic range to accommodate the expected variations of echo signal power. They must also have a low noise figure and a bandpass wide enough to accommodate the range of frequency associated with the echo pulse.

The IF amplifier amplifies the 60 MHz, or in some cases 30 MHz IF, and then demodulates it to remove the IF frequency and only pass on the echo envelope, as seen in Fig. 6–23. Finally, the echo information is fed down to the display unit to be mixed with other information to be displayed on the PPI.

Sea Clutter

Sea clutter on our display may extend three or four miles from the ship when the sea is rough, and is caused by a large amount of small returns from the sea waves surrounding the ship. These waves make excellent re-

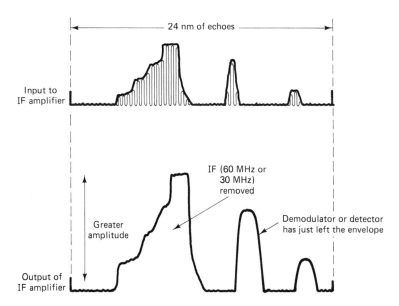

Figure 6-23 The IF amplifier's and detector's input and output waveforms.

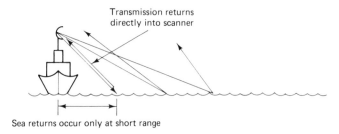

Figure 6-24   Sea echoes.

flectors for the microwave energy, and the result is to return the transmission directly back into the scanner. See Fig. 6-24.

Figure 6-25 shows how returns from the sea, clutter up the center of the PPI.

Sea clutter must be reduced, because wanted echoes near our vessel may be obscured and missed by the operator. The anticlutter sea control, which may be found on most radar control panels, reduces this clutter by decreasing the gain of the receiver's preamplifier. Wanted echoes from ships are much stronger than the clutter, so the anticlutter control is adjusted until the clutter is removed and the returns from the targets at close range are still present.

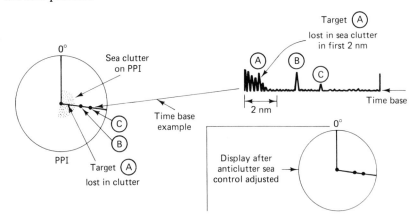

Figure 6-25   Sea clutter removal.

Automatic Frequency Control (AFC)

Figure 6-26(a) shows the magnetron frequency increasing continuously in the first half-hour after switch-on. The IF, which is the difference between the local oscillator and magnetron frequency, is continually chang-

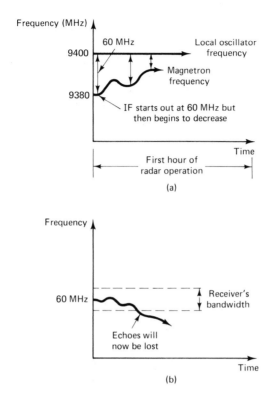

Figure 6-26 Without an AFC system. (a) Magnetron frequency drift; (b) IF drift.

ing, as shown in Fig. 6–26(b). Once the IF moves away from 60 MHz, echo information may be lost because the receiver amplifiers no longer receive the echo information being carried by our 60-MHz intermediate frequency.

Figure 6–27(a) again shows our magnetron frequency varying in the first half-hour of operation. However, with automatic frequency control the local oscillator's frequency is changed to follow the magnetron frequency change. The result is a constant 60-MHz difference between the magnetron and local oscillator, and therefore any drift by either the local oscillator or magnetron will not affect our 60-MHz intermediate frequency.

Some of the causes for IF shift are

1. Temperature changes in both magnetron and local oscillator cause changes in IF, particularly during the first half-hour of operation.
2. The power supply voltage for the AFC circuit must be very accurate with little fluctuation, as this circuit supplies and drives the local oscillator with a control voltage which controls the local oscillator's fre-

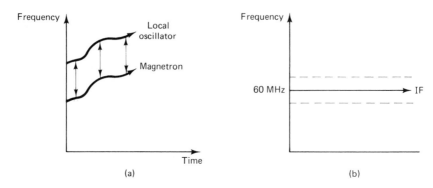

Figure 6-27 With an AFC system. (a) Local oscillator is changed to follow magnetron frequency so as to maintain IF constant; (b) IF or difference between magnetron and L.O. frequencies remains constant, so IF remains the same.

quency. Any fluctuations in the control voltage will cause an undesired change in the local oscillator's frequency, and therefore loss of IF.

The advantages of an AFC system

1. The local oscillator is tuned automatically and therefore frees the operator from continually retuning the receiver.
2. In the past local oscillators had very rigid stability requirements to ensure that they remained at their specified frequency. Now, by just changing a control voltage you can vary the local oscillator's frequency to whatever value is desired.
3. If we are assured that the IF is going to remain permanently at 60 MHz, the bandwidth of our amplifiers can be reduced to gain a better signal-to-noise ratio.

REVIEW QUESTIONS FOR SECTION 3

Refer to Fig. 6-22 for the following four questions:
1. The TCMR receiver uses an intermediate frequency of:
   (a) 30 MHz
   (b) 25 MHz
   (c) 50 MHz
   (d) 60 MHz

2. The pulse-length control to the pre-amp adjusts the amplifier's bandwidth to provide:
   (a) Short-pulse transmission, large bandwidth
   (b) Long-pulse transmission, small bandwidth
   (c) Short-pulse transmission, small bandwidth
   (d) Both (a) and (b)
   (e) Both (b) and (c)

3. The local oscillator's frequency is
   (a) 60 MHz above the echo frequency
   (b) 60 MHz below the echo frequency
   (c) 30 MHz above the echo frequency
   (d) None of the above

4. Briefly describe the function of the detector stage.

5. Sea clutter returns occur
   (a) Due to reflections from rain clouds
   (b) At short ranges
   (c) Due to land reflections
   (d) None of the above

6. Give two reasons why the IF frequency can shift above and below 60 MHz.
   (a) ――――――――――――
   (b) ――――――――――――

7. AFC is an abbreviation for ―――――――――― .

8. Give two advantages of an AFC system over manual tuning.
   (a) ――――――――――
   (b) ――――――――――――

9. An AFC system produces a control voltage to control the ――――――――――― frequency.
   (a) Magnetrons
   (b) Local oscillators
   (c) PRF generator
   (d) Wobbulator

10. Briefly describe why it is important to maintain the IF at 60 MHz.

SECTION 4 DISPLAY UNITS

Display Units in General

The display unit comprises all the electronic circuitry necessary to display the echo information in plan form on the CRT screen, as seen in Fig. 6–28. It has two main inputs:

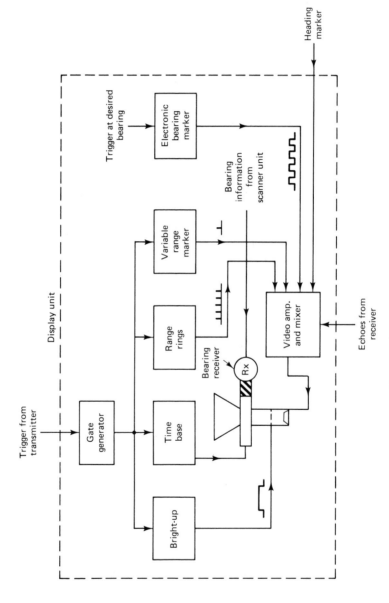

Figure 6–28 Display unit block diagram.

1. A trigger from the transmitter that initiates the gate generator waveform
2. Amplified echoes from the superheterodyne receiver.

Let us just briefly review some of the main blocks in a typical marine radar. As already just stated, the trigger from the transceiver initiates the gate generator, which produces an output waveform that switches on and off the bright-up, time-base, range rings, and range marker.

The bright-up circuit feeds a positive-going waveform to the grid of the cathode ray tube. This will increase the intensity of the electron beam travelling from the cathode to the face of CRT. A brilliance control on the radar's control panel will adjust the amplitude of this positive waveform, and therefore will change the intensity of the electron beam; consequently, the brilliance of the scene painted on the screen will be increased.

The time-base circuit produces a sawtooth, linearly rising, current waveform which is fed via slip rings to the rotating deflection coils mounted around the neck of the cathode ray tube. The electron beam from the cathode travels up and hits the center of the CRT screen—our ship's position. The sawtooth current waveform is passed through the deflection coil, which sets up a magnetic field to deflect the electron beam from the center to the edge of the screen. The direction in which the beam is deflected is controlled by the antenna, which feeds down bearing information (the bearing at which it is transmitting and receiving echoes) to control the position of the deflection coils around the neck of the CRT and therefore the direction in which the beam is deflected. The deflection coils are told to rotate at exactly the same speed as the scanner to ensure that echoes received from one bearing are displayed on the sweep at the same bearing.

Range rings are concentric rings spaced at equal intervals and allow the operator to quickly approximate target ranges.

The range marker is a variable range ring that can be placed at any range and therefore over any target, and the range of the range marker can be read from a display near the control panel.

When the scanner's beam is facing dead ahead, a switch closes in the scanner turning unit which, via a circuit in the video mixer and amplifier, causes a bright-up of one or two time-bases. This produces a bright radial line on the PPI, indicating the direction of the ship's head.

Range rings, range marker, echoes from the receiver, and a heading marker trigger are mixed and amplified in the video amplifier and mixer, which then passes the composite negative-going video to the cathode of the CRT.

Gate Generator

Figure 6–29 shows the trigger from the PRF generator in the transceiver starting the gate generator via a delay line. The start of the gate generator waveform and therefore the start of the time-base and sweep are delayed to ensure that a zero range echo will appear at the center of the PPI. To further explain this point of synchronization, let us examine the time-related waveforms illustrated in Fig. 6–30.

The display unit sync or trigger pulse informs the gate generator that transmission has begun and it should produce a gate generator waveform to turn on all its controlled circuits.

Let us imagine a seagull sitting on our scanner. This is probably hard to conceive, as our scanner is spinning around at a rate of approximately 28 revolutions per minute, and this environment would probably not be that attractive to a passing seagull. But for ease of explanation we will imagine a seagull on our scanner and therefore at zero range.

Our transmitted pulse at the magnetron's output does not appear immediately at the scanner, due to the propagation delay inherent in the waveguide run from the transceiver to the scanner. Once reaching the scanner, our pulse will strike the seagull and return a small portion of the transmitted power back to the receiver. The echo is subject to the same propagation delay between the scanner and the receiver. Once the echo is amplified and demodulated, it is then passed to the display unit, where it is mixed and finally appears as a negative-going echo at the cathode of the CRT. If the

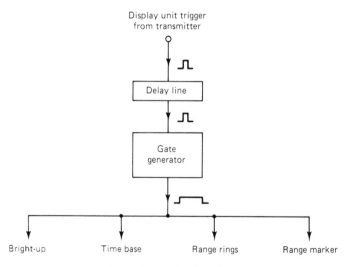

Figure 6-29   The delay line.

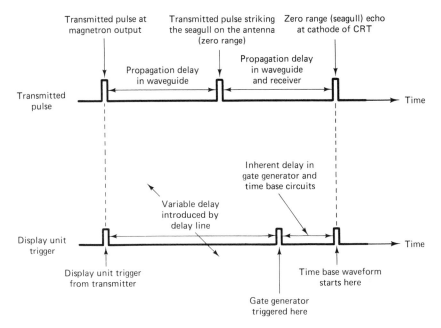

Figure 6–30   Variable delay ensuring that the time base starts at the same time that a zero range echo reaches the cathode of the CRT.

time-base began the moment we transmitted our microwave pulse of energy from the magnetron, the sweep would begin too early, as our seagull echo, which is delayed by the transmit-and-receive waveguide propagation delay and should be displayed in the center of the screen (because it is at a range of 0 nm), would appear at some range other than 0 nm. It is therefore imperative that the start of the time-base be delayed to ensure that zero range echoes occur at the center of the screen.

Figure 6–30 shows that the display trigger pulse from the transceiver is delayed before triggering the gate generator. An inherent delay in the time-base circuits means that the time-base waveform occurs a small amount of time after the gate generator signals it to, but the end result is that the time-base scan begins the moment a zero range echo would appear at the cathode of the CRT.

Time-Base

Ideally the deflection coils mounted round the neck of the CRT should be completely inductive. However, the deflection coil, like any other coil, contains both inductance and a small value of resistance. The problem we are faced with is to decide what voltage waveform would be needed to pass a ramp current waveform and produce a linearly increasing magnetic field to deflect the spot at a linear rate from the center to the edge of the CRT.

Figure 6–31 illustrates a typical deflection coil, which displays an inductance value of 10 millihenries and a resistance of 300 ohms. If the maximum deflection current (the deflection current needed to deflect the spot to the edge of the CRT) is, for example, 100 milliamps. We can calculate the voltage waveform required to produce our linear current ramp of 0 to 100 milliamps.

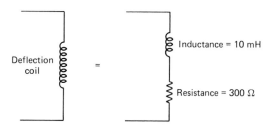

Figure 6–31   The deflection coil.

Example 1. 1-nm Range (Figure 6–32)

On the 1-nm range, the time-base current waveform goes from 0 to 100 milliamps in 12.5 μs. The voltage waveform needed to pass this linear current through a resistor is a voltage ramp which reaches a maximum voltage value of:

$$\text{Voltage} = \text{current} \times \text{resistance}$$
$$= 100 \text{ mA} \times 30 \text{ ohms}$$
$$= 30 \text{ volts}$$

A square wave will be needed for our inductance, the peak of which will be equal to:

$$\text{Voltage for inductor} = \text{inductance} \times \frac{\text{rate of change of current}}{\text{rate of change of time}}$$

$$= 10 \text{ mH} \times \frac{100 \text{ mA}}{12.5 \text{ }\mu\text{s}}$$
$$= 10 \text{ mH} \times 8 \text{ kiloamps per second}$$
$$= 80 \text{ volts}$$

The combined voltage waveform has both the ramp to pass the linear current through the resistance of the coil and the step or square wave to pass the linear current through the inductance of the coil. The voltage waveform is referred to as a *trapezoidal waveform* (from the word *trapezium*), as it has two parallel and two nonparallel sides.

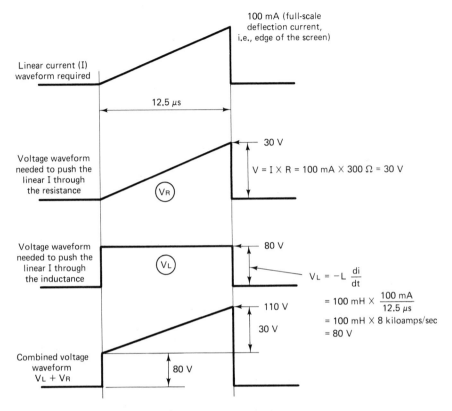

Figure 6-32 One-nautical-mile range.

If this voltage waveform were applied to a deflection coil of 10 mH and 30 ohms, a linear current would be passed through the coil from 0 to 100 mA in 12.5 μs.

Example 2. 12-nm Range (Figure 6-33)

On the 12-nm range, the time-base lasts for 150 μs. (Time-base = range in use × 12.5 μs.) Let us proceed to calculate the trapezoidal voltage waveform needed for the same deflection coils.

The ramp voltage for the resistance does not change; however, the voltage step for the inductance has changed because the rate of change of current with respect to time calculates out to be 0.6 kiloamps per second.

Figure 6-34 illustratres the two range examples just covered, and how they appear when alongside each other.

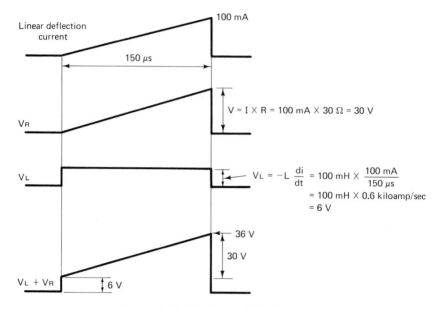

Figure 6–33   Twelve-nautical-mile range.

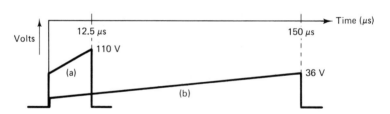

Figure 6–34   Trapezoidal voltage waveforms for (a) 1-nm range; (b) 12-nm range.

Bright-Up

The beam should be visible on the PPI only during the outward sweep of the time-base to the screen edge; otherwise:

1. Echoes returning during the flyback period (when the time-base is returning from the edge to the center of the screen) would cause a paint on the screen at the incorrect range, as illustrated in Fig. 6–35.

2. If we did not gate bright-up but left the CRT in a continuous sensitive state, then during the dead time, the stationary beam would burn the center of the screen.

3. Flyback would be seen, as shown in Fig. 6–35

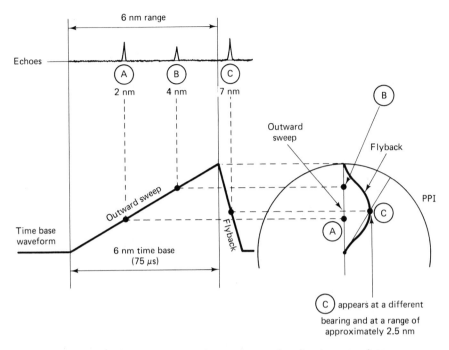

Figure 6–35 Effects of brightening up the complete time-base waveform.

Brilliance Correction

As shown in Fig. 6–36 brilliance tends to fall off at the outer edge of the PPI because the linear velocity is greater than at the center of the screen,

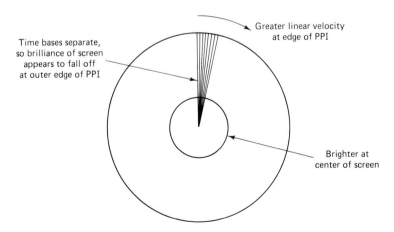

Figure 6–36 Uneven brilliance on PPI.

and the sweeps separate. A greater separation between the time-base traces means a reduction in brilliance, which in some way must be corrected. This can be compensated for by using a brightening pulse of trapezoidal shape which causes a gradual increase in brilliance to offset the fall-off. See Fig. 6–37.

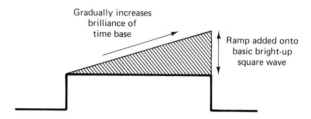

Figure 6–37   Trapezoidal or swept brilliance bright-up waveform.

Electronic Bearing Marker (EBM)

The function of the bearing marker system is to produce a dotted line of equal mark and space ratio from the center of the CRT to the edge at any required angle, and to provide a digital readout of the angle by means of a digital display. Figure 6–38 shows that the electronic bearing marker control, when adjusted, will achieve two functions:

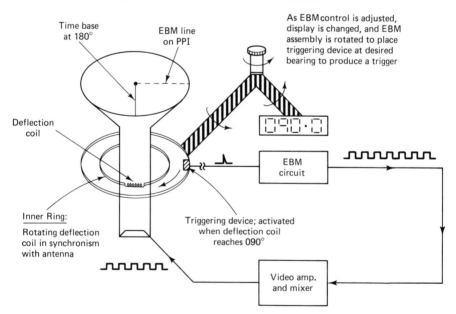

Figure 6–38   EBM.

1. It rotates an assembly around the neck of the CRT (separate from the deflection coils). This assembly, depending upon its position, will cause a trigger to be sent to the EBM circuit informing it to produce a dotted bearing marker line at that bearing. The trigger is activated by the time-base scan when it reaches that bearing.
2. The EBM control is also geared to the digital display and will automatically increment it from 000.0 to 359.9 degrees.

Figure 6–39 shows the PPI set up with the EBM and VRM being used to acquire the range and bearing of a target.

Heading Marker

To indicate the heading of the ship, a switch in the aerial turning unit closes every time the center of the radar beam cuts the fore and aft line of the ship (that is, whenever the scanner points dead ahead). This switch initiates a circuit in the display unit which causes a negative-going rectangular

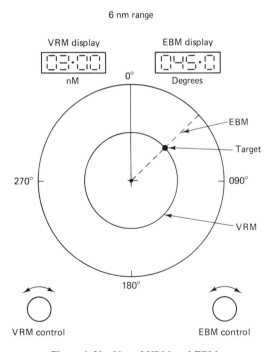

Figure 6–39   Use of VRM and EBM.

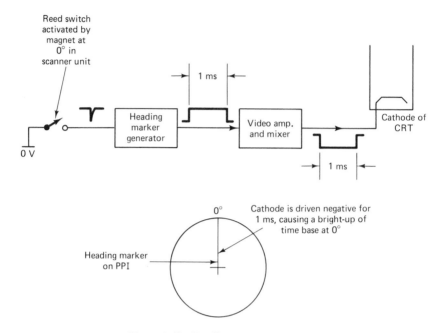

Figure 6–40   Heading marker generation.

pulse (of approximately 1 millisecond) to be applied to the cathode of the CRT, causing a bright-up of one or two time-bases. See Fig. 6–40.

Performance Monitors

Performance monitors are used to show whether the radar is working at full efficiency or not. Let us take an example of a system that incorporates two separate performance monitor checks (see Fig. 6–41).

1. A cavity resonator is mounted on our waveguide run up to the antenna. When the magnetron fires, it causes oscillations within this cavity which are sent down to the receiver and then on to the display unit to be displayed on the PPI. If the result on the PPI is correct, we would have checked the complete radar system excluding the waveguide run above the cavity and the scanner.

2. A neon power monitor is mounted near and points to the scanner. The RF transmission out of the scanner is sensed by the neon lamp, and then is fed to an amplifier and power monitor meter which gives a relative value of power being transmitted from the scanner.

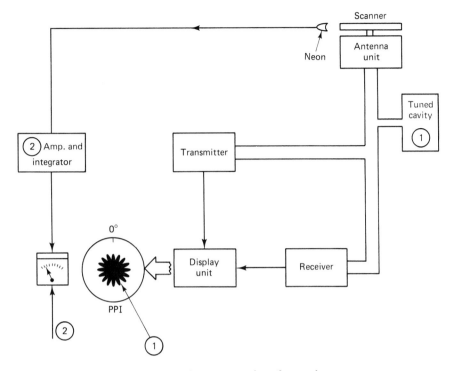

Figure 6-41   Two performance monitors for a radar system.

Let us now proceed to investigate these two performance monitors in a little more detail.

Power monitor (Fig. 6-42)   The power monitor employs a neon lamp mounted on an arm usually behind the scanner unit, so that during every revolution of the scanner, the beam sweeps through and hits the neon lamp. When the power monitor is selected by a control on the display unit control panel, the neon bulb is partially ionized and the RF energy from

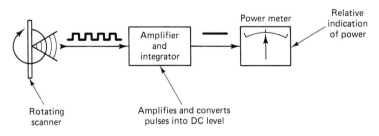

Figure 6-42   The power monitor.

the scanner increases the amount of current flow in the neon. The result is to send pulses (one pulse every 360 degrees) down to the display unit; the more powerful the scanner's beam, the larger the amount of current pulse produced. In the display unit, these pulses are amplified and then integrated to produce a steady DC voltage, which is applied to a meter mounted on the display unit.

Cavity Resonator.   Figure 6–43 illustrates a cavity connected to the waveguide run going from transceiver to antenna. When the cavity resonator performance monitor is selected on the display unit control panel, the DC solenoid and motor are energized. The DC solenoid moves the plunger back, so it is now touching the off-centered cam connected to the DC motor. As the motor rotates the off-centered cam, the plunger is moved back and forth in the cavity, continually changing its dimensions and therefore the tuned frequency. The cavity is normally swept through all the possible magnetron frequencies that are available within a band of frequencies. When the plunger has tuned the cavity to our magnetron frequency, energy from the magnetron, which is coupled into the cavity via the coupling hole, causes oscillations which travel out of the coupling hole into the waveguide.

The receiver picks up the signal, and the result on the PPI is a spoked

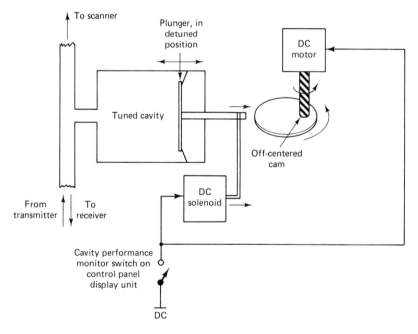

Figure 6–43   Cavity resonator system.

sun pattern. The presence of a spoke is due to the cavity and receiver's being in tune, and the absence of a spoke is caused by the cavity being taken out of tune by the plunger. See Fig. 6–44.

The cavity resonator checks our complete system except for the waveguide run above the cavity and the scanner, because if the transmitter was not operational, the cavity could not be excited into oscillation. Once oscillations occur, they are mixed, amplified, and detected by the receiver and then are fed to the display to be displayed on time-bases. Consequently, transmitter, receiver, and display unit are all operational if the correct spoked sun is obtained on the PPI.

As mentioned previously, the power monitor will check that the transmitter is radiating power up the waveguide, and out of the antenna.

Using these two performances monitor checks, the technician can analyze and isolate a fault to one of the units; for example:

Power monitor	Cavity resonator	Area of Problem
Correct	Correct	Complete radar operational
Incorrect	Incorrect	Transmitter substandard
Correct	Incorrect	(Time-base present) Receiver or video amplifier substandard
Incorrect	Correct	Loss of power in waveguide run above cavity resonator and scanner

Explanation of Previous Table

Power monitor and spoke sun incorrect. If both performance monitors are not operational, a common problem must exist. The transmitter is the only unit which will cause both performance monitors not to operate.

Power monitor correct but spoke sun incorrect. If the power monitor is giving us a correct reading on the meter, the transmitter waveguide run

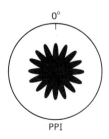

Figure 6-44 Spoked sun from cavity resonator on PPI.

and scanner must be operational. If the time-base is present, then the display unit is operational. Our problem must exist between the oscillations coming out of the cavity and getting to the video amplifier in the display unit, that is, the receiver and video amplifier.

Power monitor incorrect, but spoke sun correct. If the spoke sun is present, then all units are checked and our problem must exist above the cavity resonator, that is, the waveguide and scanner.

Of course, one problem that has not been mentioned is that our performance monitors themselves could be substandard, and therefore lead a technician in the wrong direction.

Display Presentations

The plan position indicator (PPI) presents and displays our surrounding echo information in plan form as if we were in a helicopter above the ship looking down on the surrounding situation. This is always the case; however, the information can be displayed on the PPI as either a relative motion or true motion presentation.

Relative Motion (RM) Unstabilized: Ship's Head Up

Figure 6–45 illustrates a ship heading 45 degrees to a lighthouse. Our display on the PPI always shows the ship's heading up, and so the lighthouse appears at 045 degrees, and the land directly behind the lighthouse.

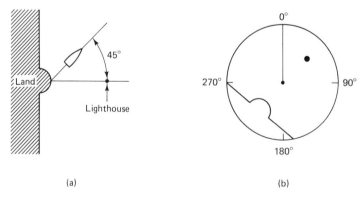

(a)   (b)

Figure 6–45   (a) Aerial view; (b) PPI display.

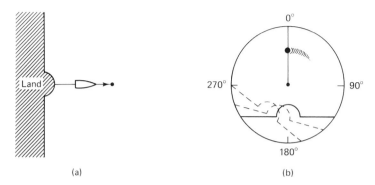

Figure 6-46 Ship course change. (a) Aerial view; (b) PPI display.

If the ship were to now change course and head directly for the light vessel, the display on the PPI would appear as illustrated in Fig. 6–46.

Advantage

1. Easily recognized situation. If a target appears dead ahead of the ship, it will be presented dead ahead on the display.

Disadvantages

1. Relative motion. Stationary targets, as in the case of the land in the above example, will move off our display as the ship proceeds on course. This is a false presentation, because in fact our ship is moving away from the land instead of the land's moving away from the ship, as displayed on the PPI.
2. Targets blur as ship changes course, as illustrated in Fig. 6–46.
3. The bearings taken of targets are not true bearings relative to north; they are bearings relative to the ship's heading. Since all charts are drafted with north up, chart comparison with our PPI display becomes difficult.

Relative or Azimuth Stabilized Display (North Up)

Figure 6–47 shows a ship heading northeast, and a similar situation to the previous display where a lighthouse is 45 degrees to the ship's heading (east). The PPI is now a stabilized north-up display, and the heading marker will move and appear at the correct heading of the ship, in this case northeast.

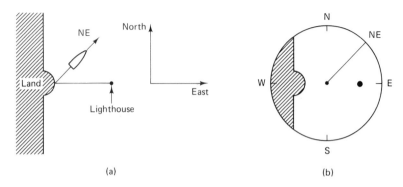

Figure 6-47 RM stabilized.

In Fig. 6-48 the ship has changed course and is now heading due east directly for the lighthouse. As can be seen on the PPI, the only change occurring is that of the heading marker, which now moves around to now point east, but all of our targets remain relative to north in their same position.

Advantages

1. Targets do not blur when the ship changes course.
2. True bearings of targets are given, thus making chart comparison a lot easier.
3. As with any relative motion display, a collision risk is easily recognized.

Disadvantages

1. Relative motion; that is, fixed targets move as you proceed on course.

True Motion Display

With a true motion presentation the position of your ship, representated by the point from which the rotating trace originates, is moved across the face of the CRT by means of off-centering circuits. These off-centering arrangements consist of a stationary square coil mounted around the rotating deflection coils. The current in the stationary coil, and hence the point from which the rotating trace originates, is controlled by the true motion unit in such a way that the movement of the trace origin across the CRT face is at a rate and direction corresponding to the ship's speed and

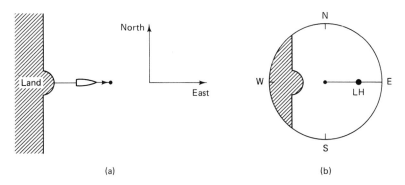

(a) (b)

Figure 6-48 Course change on a RM stabilized display.

course, respectively, thus resulting in the true motion of all targets on the PPI; that is, echoes from fixed targets remain stationary, and echoes from moving targets proceed along their course.

Comparison of Relative and True Motion Presentations

Figure 6–49(a) is a relative motion, north-up display and Fig. 6–49(b) is a true motion presentation. To fully understand a true motion display we will compare the two displays and see how the true motion display actually gives the more accurate presentation, as if you could hover permanently over the surrounding area around your ship, but the relative motion display has its own advantages over true motion.

The arrows show the direction in which the echoes are travelling on the display, which can be different from their actual direction, as will be seen when we discuss the following examples.

Targets. In Figs. 6–49(a) and 6–49(b) *A* is a stationary target and is presented as a stationary object on the TM display; on the RM it is seen to travel in the opposite direction at a speed equal to the ship's speed.

Target *B* is on the same course as ourselves, but travelling at a slow speed; that is, we are overtaking. On the TM display, the correct or true presentation is seen, but the RM display shows *B* travelling in the opposite direction—that is, although *B* is travelling in the same direction, we are picking up the return at a constantly decreasing range because we are overtaking, and so on the display he appears to travel in the opposite direction.

Target *C* is on a collision course and the TM display shows exactly how our ship and target *C* would actually appear if viewed from a heli-

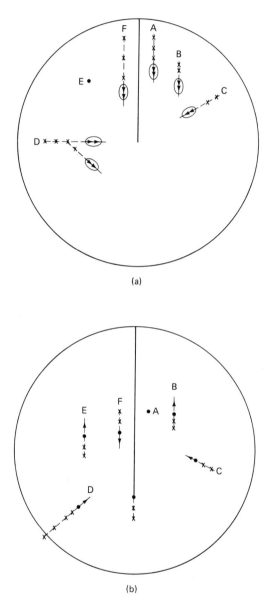

(a)

(b)

Figure 6-49 (a) Relative motion (RM) stabilized (north-up) display; (b) true motion (TM) display.

copter. The RM display easily indicates to the operator that a collision is developing with target C.

Target D is proceeding on a course of due east, and has slowed down without a change of course to avoid collision with our ship. On the TM display this appears exactly as previously stated, but on the RM display it appears that target D has actually altered course. This is not a true presentation; target D is just being picked up at an increasing distance, and that is why it appears to have changed course on the RM display.

Target E is on the same course as our ship, and is proceeding at exactly the same speed. On the TM display a true replica of the situation is displayed, but on the RM display target E appears as a stationary target because it is being picked up at exactly the same bearing and range from our ship.

Target F is on a reciprocal course to ourselves, and proceeding at a similar speed. The TM display shows the true presentation, while the RM display makes target F appear to be travelling at a rapid speed. As we remain stationary in the center of the screen of a RM display, target F is travelling in the opposite direction at a rate equal to the sum of its ship's speed and ours.

Each method of presentation has its own particular advantages: Relative motion shows target B getting nearer although it is actually heading away, while on the true motion display it is not readily apparent that target C is on a collision course. Some ships that have two radars find it an advantage to have one presenting true motion while the other displays relative motion.

TCMR Display Unit

The TCMR display unit contains a power supply to develop the DC voltages required for the whole system, the main display board, which will be discussed first, the interference rejection unit, and the range marker/delay unit.

Figure 6–50 illustrates the TCMR display unit block diagram.

Let us first examine in a little more detail the bearing transmission blocks in the display unit, and the associated components in the scanner unit. See Fig. 6–51.

The bearing circuit receives a train of pulses, which are generated in the scanner unit, and drives a stepper motor, which rotates the deflection coil around the neck of the CRT, producing a rotating time-base.

A four-pole magnet is mechanically coupled to the antenna via a gearbox. This four-pole magnet rotates at approximately 110 revolutions per

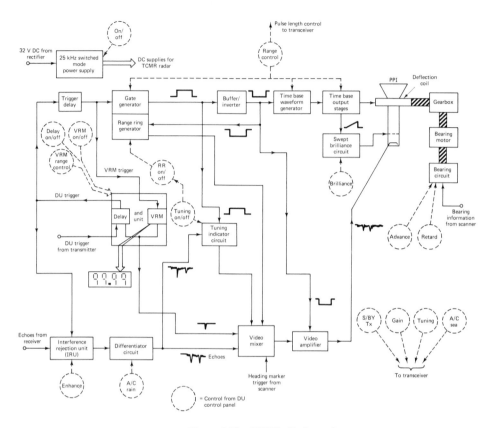

Figure 6-50   TCMR display unit.

one scanner revolution; each time the four-pole magnet completes one revolution, four pulses are induced in an inductor and are then coupled to the display unit. Consequently, in one complete scanner revolution, we have a total of:

$$110 \text{ revolutions} \times 4 \text{ pulses per revolution} = 440 \text{ pulses}$$

440 pulses are received down in the display unit for one revolution of the scanner through 360 degrees. These 440 pulses drive a small stepper motor, which via a gearbox will drive the deflection coils once around the neck of the CRT.

   Generally, the rotation of the time-base is in step with the scanner, but when the TCMR is shut down after use, the deflection coil will stop almost instantaneously; however, the scanner slowly comes to a halt, resulting in the scanner and deflection coils at different bearings. When the radar is turned back on, the deflection coils are kept at exactly the same

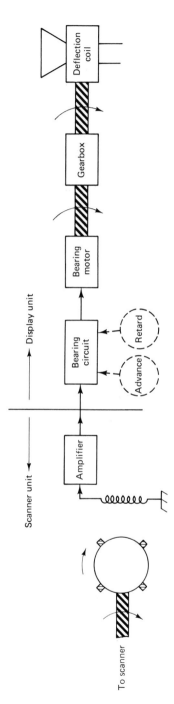

Figure 6–51 Bearing transmission system for TCMR.

speed as the scanner, but, for example, if a 90-degree difference was present between the two before switch-on, then this 90-degree difference will still exist and be maintained. It can be recognized on the PPI because the heading marker, which is normally at 0 degrees, is now at, for example, 270 degrees, as illustrated in Fig. 6–52.

This means that when the radar was switched off, the deflection coils stopped at 270 degrees, and the scanner drifted on another 90 degrees to 360 degrees. Now that our radar has been turned back on, the scanner is leading the deflection coils by 90 degrees. Consequently, when the heading marker switch closes, telling the display unit to brighten up one or two time-bases, the deflection coils are pointing to 270 degrees, and so the time-bases are brightened up at that bearing. The echo from the target, which is at a relative bearing of 0 degrees, will also be placed at 270 degrees because of this error.

There must obviously be a way of achieving positional synchronization between scanner and deflection coils, and the TCMR system has two methods:

1. The retard switch on the control panel, when depressed, inhibits the flow of all pulses, and therefore stops the rotation of the deflection coil and consequently the trace, to wait for the scanner. This is a coarse adjustment.
2. The advance switch, when depressed, enables ON an oscillator, which then feeds additional pulses to the bearing motor, so it speeds the deflection coil to catch up with the scanner. This is a fine adjustment.

Either or both of these methods can be used to realign the heading marker to 0 degrees for a relative motion unstabilized (head-up) display.

The display unit trigger, once it arrives on the main display board (from the transceiver), is fed to trigger the VRM/delay unit (which will be

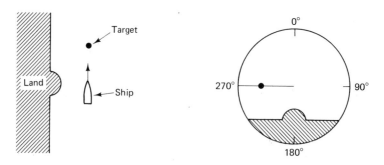

Figure 6–52   Scanner and deflection coil misalignment.

explained later), and then to a trigger delay circuit, which will delay the trigger and therefore the start of the time-base until the moment a zero range echo would appear at the cathode of the CRT. After the delay the trigger initiates both the range marker and the gate generator, whose output waveform width is controlled by the range control input set by the operator. The range-ring calibrator circuit, which is part of the gate generator, is driven by a negative gate waveform via a buffer/inverter. The range rings will be produced if the range-ring switch is on and the tuning indicator switch is off. (The reason that the tuning indicator switch must be off will be explained shortly.) The range pulses will be fed to the video mixer and then amplified before reaching the cathode of the CRT to cause evenly spaced dots on every time-base and will therefore appear as concentric rings on the PPI.

The positive gate waveform is buffered and inverted, and then is fed to the time-base waveform generator and output stages, which generate a trapezoidal voltage waveform to produce a linear current through the deflection coil. The input from the range selector control to the time-base blocks determines the steepness of the ramp waveform, and the gate generator's waveform controls the duration of this time-base waveform.

A sawtooth waveform derived from the time-base sawtooth is applied to the grid of the CRT via the swept brilliance circuit. This maintains a constant brilliance from the center of the PPI to the screen edge. Intensity of the display can be controlled by a brilliance control on the control panel.

The echoes from targets are fed to the display unit via the interference rejection unit, and are differentiated when the anticlutter rain control is engaged.

Because of the small wavelength at microwave frequencies, rain appears large and therefore will return some of the transmitted power back to the radar and appear as seen in Fig. 6–53.

The rain clutter can be broken up by differentiation, so instead of one large echo appearng on the PPI, the differentiated rain will appear as tiny dots, and the echo that was previously hidden is now a recognizable target.

The echoes are then fed to the video mixer, where they are mixed and then fed to the cathode of the CRT via the video amplifier.

The echoes are also fed into the tuning indicator circuit, which when switched on will produce a pulse for every time-base. The range of this pulse will be determined by the amount of echoes received at the input of the tuning indicator circuit. That is, a tuning control on the display unit feeds a control voltage to the local oscillator to vary its frequency manually; once the IF is at 60 MHz, the AFC circuit locks on to maintain the IF at 60 MHz by automatically varying the L.O.'s tuning voltage. When the operator is adjusting this tuning control to ensure that the echoes are at the IF frequency and therefore passed by the IF amplifier, some indication must be

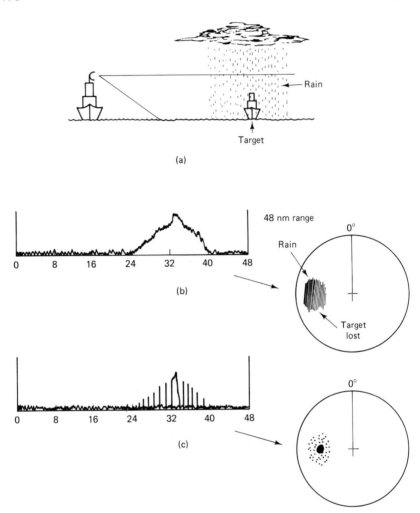

Figure 6-53 (a) Target in rain; (b) no differentiation; (c) anticlutter rain control differentiates and breaks up rain.

given to assist the operator in this tune-up. The tuning indicator circuit will produce one pulse for every time-base. This pulse will appear at a range determined by the amount of echoes received into this circuit. In other words, the better the tuning, the better the echo strength, and therefore the pulse will appear at a greater range. If a pulse is produced for every time-base, a concentric ring will be produced; the tuning control is adjusted to send this ring to a maximum range, which when achieved will mean that the echoes are now in tune with the IF amplifier. See Fig. 6-54.

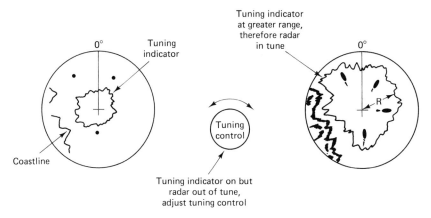

Figure 6-54 The tuning indicator ring.

The tuning indicator pulse, the range rings, the echoes, the range marker pulse, and the heading marker input, are all mixed, amplified, and then fed to the cathode of the CRT when intensity modulation of the electron beam presents all this information on the PPI. The video amplifier is gated to allow video information to the CRT only during the time-base's output sweep.

TCMR's Interference Rejection Unit (Video Processor)—IRU

When a radar is in the vicinity of other radars using the same frequency band, interference occurs and appears as a large number of bright dots scattered over the whole screen, as seen in Fig. 6–55.

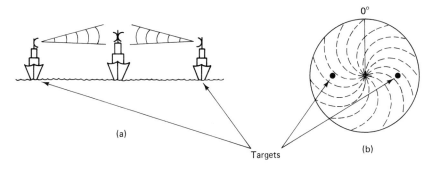

Figure 6-55 (a) Nearby ships causing (b) radar interference on PPI.

The interference rejection unit will remove most of this radar inte-
ference and also will improve the clarity of the information from the IF
amplifier before sending it to the display unit.

Interference from other ship's radars occurs at random, and will not
occur at the same range on two consecutive time-bases. The IRU will take
a current time-base (T/B 1), convert the analog echo information into digital form, and then store this time-base of echoes. See Fig. 6-56.

Time-base 2 of echo information is also converted to digital form and
then stored. The two time-bases are then compared, time-base 1 (previous)
and time-base 2 (present), and as the radar interference does not correlate

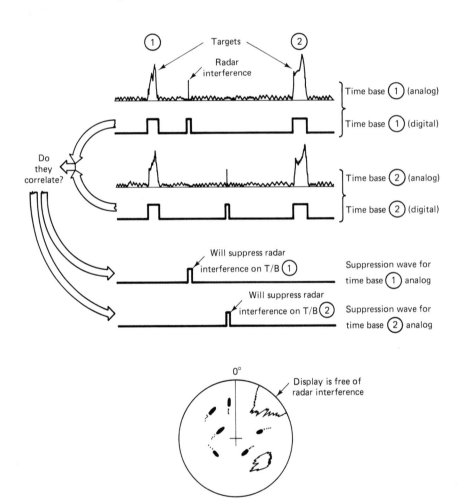

Figure 6-56   The interference rejection unit's removal of radar interference.

on either time-bases, then suppression pulses are produced to remove the interference before it can pass out of the IRU to the display unit. The target echoes 1 and 2 correlate and therefore are allowed to pass. This is a continuous on-going operation in which time-base 2 is compared with 3, 3 with 4, 4 with 5, and so on to produce a radar interference-free pciture.

An enhance switch, when engaged, will widen the width of small pulses so as to allow them to be viewed more easily by the operator. A timing circuit inhibits this enhance action for the first two miles of every time-base to prevent cluttering of the screen center.

Range Marker/Delay Unit

The function of the range-marker section of this unit is to produce one output pulse for every time-base at an adjustable time interval, which can be varied by the range-marker range control. The range at which the pulse is set is displayed on a digital readout in nautical miles, and one pulse per time-base appears as a movable concentric range ring. See Fig. 6–57.

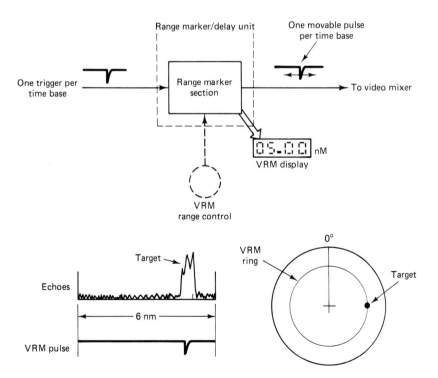

Figure 6–57   The TCMR VRM system.

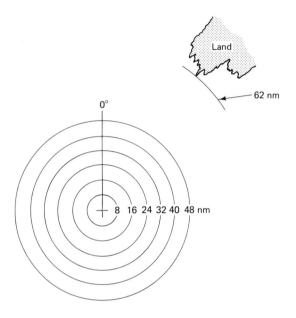

(a) Range selected = 48 nm
     VRM setting = 0 nm
     Delay button = OFF

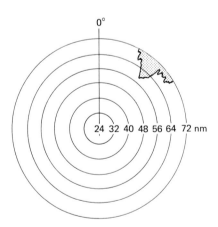

(b) Range selected = 48 nm
     VRM setting = 24 nm (triggers display unit)
     Delay button = ON

Figure 6-58   Range extension.

This variable range marker allows the operator to simply and accurately measure target range by placing the ring over the target's echo on the PPI.

The delay mode can be employed to:

1. Extend the range of the radar system.
2. Expand a specific area for examination.

1. Range Extension. Figure 6–58 shows how the delay mode can be utilized for extending the range of your radar.

The display unit is set at a range of 48 nm, which means that 6 range rings have been produced, each of which have been spaced at 8-nm intervals. Land exists at 62 nm, but cannot be seen on our radar screen because it is beyond the range selected.

If the range marker is set to 24 nm, and the delay button on the range marker/delay unit is depressed, the display is no longer triggered by a trigger from the transceiver at 0 nm, but instead is triggered by the range-marker pulse at 24 nm. This results in our gate, time-base, and everything else starting at a range of 24 nm, and as we are on the 48-nm range, our time-base will now extend 48 nm away from our 24-nm start; that is, to 72 nm. The radar range has now been extended to 72 nm, but it must be realized that targets at a range less than the set display range will not appear on the PPI.

2. Range Expansion. The range expansion technique is illustrated in Fig. 6–59. There are three targets that are grouped between 3 nm and 6 nm; as can be seen, by setting the range-marker delay to 3 nm and the range

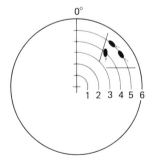

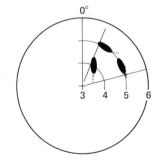

(a) Range scale = 6 nm
    VRM setting = 0 nm
    Delay button = OFF

(b) Range scale = 3 nm
    VRM setting = 3 nm (triggers display unit)
    Delay button = ON

Figure 6–59   Range expansion.

to 3 nm, we can now closely examine the three targets because our time-base starts at 3 nm, and extends 3 nm, which is the exact area of our three targets.

25 kHz Switched Mode Power Supply

The power supply chassis is mounted in the display unit, and produces all the necessary DC voltages needed for the operation of the radar set.

This power supply's drive waveforms operate at a frequency of 25 kHz. Of the various voltages produced at the output, one of them, which will represent all of them, is fed back to a control board and analyzed for variations by voltage-sensing circuits. The resulting error signal, which is proportional to the output voltage variations, varies the width of the drive waveforms going through to develop the final DC voltages. This closed-loop feedback system will regulate the DC output voltages, keeping them at the desired values.

Switching power supplies get their name because they switch through drive only when it is needed to develop the output DC voltages. The old design of power supplies was to take whatever was given and dissipate power until you gained the desired voltage. This used to work fine and also kept you warm during the winter months, but was obviuosly not very efficient. The switch mode power supplies, however, enable you, instead of just accepting what is delivered and convertng it to what you want, to control what is delivered by going back to correct the source.

REVIEW QUESTIONS FOR SECTION 4

1. The signal arriving from the transmitter to the display unit is the
   (a) Trigger
   (b) Echoes
   (c) Heading marker
   (d) Bearing information
2. The signal(s) arriving from the receiver to the display unit is the
   (a) Trigger
   (b) Echoes
   (c) Heading marker
   (d) Bearing information

3. The time-base waveform:
   (a) Rotates the deflection coils around the CRT.
   (b) Causes the beam to sweep from the center to the edge of the screen
   (c) Both (a) and (b)
   (d) None of the above

4. Give three reasons why the bright-up waveform ensures that the beam in visible only during the outward sweep from the center to the edge of the screen.
   (a) _____
   (b) _____
   (c) _____

5. EBM is an abbreviation for _____ .

6. Why is the EBM a dotted line?

7. Briefly describe the function of the radar's performance monitor.

8. A relative-motion unstabilized display has three disadvantages, which are
   (a) _____
   (b) _____
   (c) _____

Refer to Fig. 6–50 for the following questions:

9. The advance and retard switches for the bearing circuit are needed due to:
   (a) The gate generator's being unstable
   (b) Scanner and deflection coil misalignment
   (c) Inaccuracies in the power supply
   (d) None of the above

10. The advance switch will:
   (a) Never be used
   (b) Stop the deflection coils
   (c) Adjust the length of the time-bases
   (d) Add additional pulses to the bearing information

11. Clutter on the PPI due to rain can be reduced by use of the:
   (a) Enhance switch
   (b) Delay switch
   (c) Radar on/off switch
   (d) A/C rain control

12. The TCMR's IRU removes:
   (a) Rain clutter
   (b) Sea clutter
   (c) Radar interference
   (d) All of the above

13. The delay unit section of the VRM/delay unit can be used to:
   (a) Extend the range of the radar
   (b) Reduce radar interference
   (c) Expand an area for examination
   (d) Both (a) and (b)
   (e) Both (a) and (c)

14. Briefly describe the advantage of a switch-mode power supply over a standard power supply.

15. List the five inputs to the video mixer, which are combined to form composite video:

    (a) ——————————————

    (b) ——————————————

    (c) ——————————————

    (d) ——————————————

    (e) ——————————————

SUMMARY

1. The scanner, transmitter, receiver, and display unit are the four main sections of a radar system.

2. The scanner unit houses the bearing transmitter, the scanner, the heading marker circuit, and the scanner turning motor.

3. The transmitter houses the trigger circuit, the modulator, and the magnetron.

4. The receiver holds the T/R switch, the mixer and local oscillator, the IF amplifier and detector, and the automatic frequency control circuit.

5. The display unit receives a trigger from the transmitter, echoes from the receiver, and bearing information from the scanner. It holds the necessary circuitry to display the echoes and synthetic video on a plan position indicator.

7

SATELLITE COMMUNICATIONS

Objectives

After completing this chapter, you will be able to:

1. Identify the main satellite organizations.

2. List the names and functions of the satellites in the INMARSAT, INTELSAT, and U.S. regional groups.

3. Explain:

   a. Frequency division multiplexing

   b. Time division multiplexing

   c. Pulse modulation

   d. Shift keying

4. Describe the function of the teleprinter.

5. Explain the frequencies, signal formats, and full operations of a marine satellite communication system.

Messages of any, and all, types are passing into and out of space. Orbiting geostationary satellites operating at frequencies between 3.7 to 17 GHz are becoming a very important form of communication.

SATELLITE CORPORATIONS

The communications satellite corporation COMSAT was a conglomerate set up in 1963. INTELSAT followed with more than six synchronous-orbit satellites 22,240 miles above the equator over the Atlantic, Pacific, and Indian Oceans. INTELSAT handles about two-thirds of all transoceanic communication traffic, and is owned by 105 nations, with COMSAT as the U.S. representative. Almost all countries, except the Russian Warsaw Pact group, are owners, and almost all have active government-owned earth stations.

MARISAT is owned by COMSAT and three other international carriers, and provides satellite communications for commercial shipping and the U.S. fleet by utilizing three satellites over the Indian, Atlantic, and Pacific Oceans. The international marine satellite organization (INMARSAT), of which thirty nations are members, is another marine satellite group set up because MARISAT was becoming badly congested with an increasing number of marine satellite communication terminals.

COMSTAR, which is part of the COMSAT organization, has four satellites, each of which can handle 18,000 telephone conversations, and all of which are leased by AT&T.

All of this breaks down to two major satellite communication organizations:

1. INTELSAT, whose satellites provide point-to-point transoceanic communications.
2. INMARSAT, which provides communications for ships at sea.

SATELLITES

The satellite receives the signals transmitted to it by the earth station. It then amplifies and sends this information to another earth station on a new carrier frequency. A frequency difference of about 2 GHz prevents interference between the uplink and downlink transmissions. For example, all geostationary satellites above the United States operate in one of the following three bands:

Old Band	Uplink	Downlink	Orbit Separation
C	6 GHz	4 GHz	4 degrees
Ku	14 GHz	12 GHz	3 degrees
K	17 GHz	12 GHz	Not assigned

Let us now take a closer look at the INTELSAT, INMARSAT, and U.S. regional satellites.

1. INTELSAT Satellites

In 1965, INTELSAT 1, or the "Early Bird," was launched and placed over the Atlantic Ocean and was used to provide 66 simultaneous transatlantic telephone conversations. There are now more than 12 INTELSAT satellites over all three oceans, offering:

a. 12,500 telephone circuits.
b. Two television channels per satellite.

The INTELSAT V1 satellites, which are expected to be launched in the late 1980s, will provide:

a. 20,000 telephone circuits per satellite.
b. International and domestic TV services.

An expected 500 earth stations in 150 countries will make use of this system.

Figure 7-1 shows an INTELSAT V satellite that is currently being used for global point-to-point satellite communications. These satellites use:

UPLINK FREQUENCY	*DOWNLINK FREQUENCY*
5.925–6.425 GHz	3.7–4.2 GHz

INTELSAT V uses frequency modulation (FM), frequency division multiplexing (FDM), and time division multiplexing (TDM). FDM and TDM are covered later in this chapter.

2. INMARSAT Satellites

In 1976, ship-to-shore and shore-to-ship long distance communication was no longer limited to only HF radio, but could now be beamed to geo-

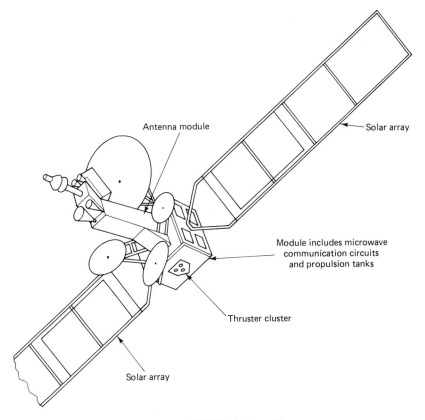

Figure 7-1   INTELSAT V satellite.

stationary satellites providing high-quality telephony, and telex/telegraphy communication.

This initial MARISAT system had three satellites over each ocean with an uplink frequency range of 1.5 to 1.6 GHz, and a 6 to 4 GHz downlink. Figure 7-2 shows a MARISAT satellite.

These five-year-design-life satellites quickly begin to feel congestion, and the International Marine Consultative Organization (IMCO) commissioned a group of experts to recommend a new marine satellite organization. INMARSAT was born in 1979 with the U.S. (COMSAT) as one of the largest shareholders. There are now 20 INMARSAT earth stations with antennas 13 meters in diameter, and with additional communication capacity being leased from INTELSAT V satellites, the congestion has been almost completely relieved.

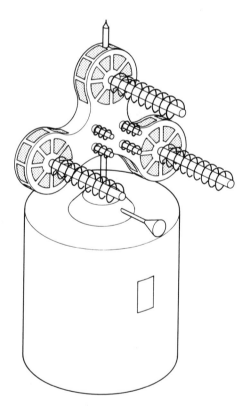

Figure 7-2   A MARISAT satellite.

3. U.S. Regional Satellites

a. SATCOMS IR, IIR, III and IV.   are owned by RCA, and each has 24 channels which can handle:

(1) 3000 telephone conversations via AMSSB Techniques through 8.5W solid state transponders.
(2) An FM/color television transmission.
(3) 60 megabits (1 megabit is a million binary digits) of computer data per second.

Power is provided by two rotating solar arrays, each of which has a solar area of 75 square feet. The second generation SATCOM's (IR and IIR) have 90 square feet of solar array, and this extension provides a power increase from 5.5 to 8.5 watts. Figure 7–3 illustrates the RCA SATCOM satellite.

b. GSTAR (Ku Band).   Owned by GTE Corporation, these satellites digitize communication, using 16 transponders and time division multiplexing (TDM). A transponder is a device that receives information from a source (the uplink), and then transmits exactly the same information at another frequency (the downlink).

This new series of satellites will serve a variety of customers with a range of voice, video, and data services.

Figure 7–4 illustrates the GTE GSTAR satellite.

c. WESTAR®.   Western Union was the first U.S. company to place domestic satellites in orbit in 1974. The first satellite carried 12 transponders and were used for:

(1) Computer data
(2) Video
(3) Voice
(4) Facsimile

Facsimile is a process whereby a picture is scanned at the source; the information of that picture is converted to electrical signals which are then

®WESTAR is a registered service mark of the Western Union Telegraph Company.

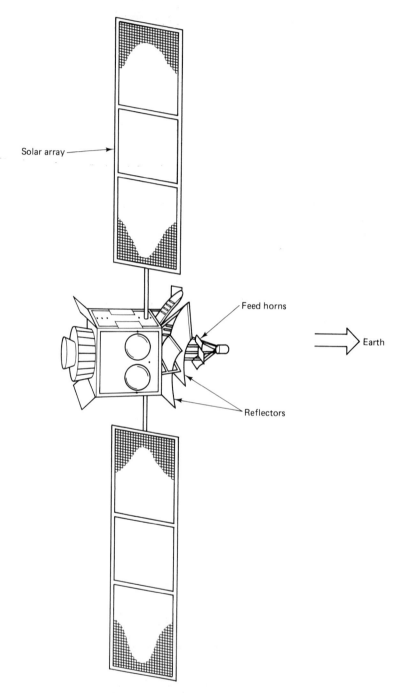

Figure 7-3   RCA SATCOM satellite.

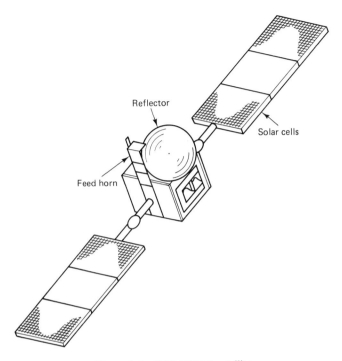

Figure 7-4   GTE GSTAR satellite.

transmitted, in our case, via satellite to produce a likeness (facsimile) of the picture at the destination.

The current WESTARS have 24 transponders and can provide:

(1) 2,400 one-way voice communications

(2) One full-color television transmission

(3) Transmission of 60 megabits of digital data per second per transponder

They are designed to last a minimum of ten years; they utilize 7.5-watt TWTs, and 800 watts of solar power is generated by the solar cells. Figure 7-5 illustrates the WESTAR 1V satellite.

d. ANIK.   Both ANIK C (Ku band) and the advanced ANIK D (K band) provide Canada with audio, video, and digital data telecommunications. ANIK D covers all of Canada, while ANIK C services Canada's more densely populated southern area.

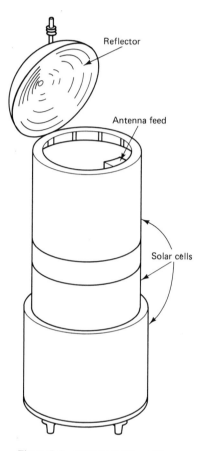

Figure 7-5　WESTAR IV satellite.

SATELLITE SIGNALS

Using microwave frequencies means that there must be a line-of-sight propagation path for the uplink and downlink. As mentioned previously, the
uplink is transmitted at a higher frequency to prevent cross talk. Between
uplink being received, and downlink being transmitted, we have a transponder within our satellite.

There are two main methods for transmitting intelligence to the satellite and then recovering it at the receiving end. Before continuing on with
the satellite signals, we shall first discuss, in general terms, multiplexing.

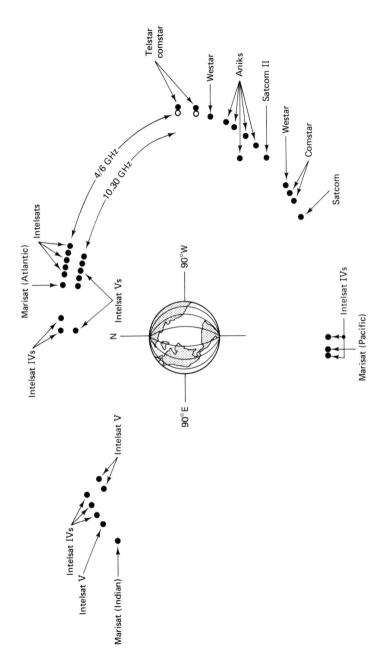

Figure 7-6 International, marine, and U.S. regional satellites.

MULTIPLEXING (many into one)

Multiplexing, in communications, means to interleave or transmit two or more messages simultaneously on the same carrier frequency.

1. Frequency Division Multiplexing (FDM)

This is a system whereby the available frequency range (bandwidth) is divided into narrow bands, each of which will be used to transmit a separate channel; that is, this process will allow two or more signals to be transmitted over a common path by sending each one over that path at a different frequency.

Figure 7-7 illustrates an example of FDM in which twelve different signals (for example, one-way telephone conversations) are multiplexed into a channel group from 60 to 108 kHz, that is, a center carrier frequency of 84 kHz with a 48-kHz bandwidth. Each telephone conversation has a 4-kHz bandwidth, and each 4-kHz signal modulates a separate carrier frequency; that is, signal 1 modulates 64-kHz, signal 2 modulates 68-kHz, and so on. The 4-kHz telephone signals on the carrier frequencies are combined and then passed on to the telephone network. At the receiving end, filters separate and pick out each telephone signal, and demodulators remove the carrier frequencies; our remaining voice signals are then passed on to the called party.

The basic 12-channel group when combined with four other 12-channel groups forms a supergroup. Supergroups are combined to make mastergroups, mastergroups form supermastergroups, and so on.

It should be noted that Fig. 7-7 is a simplified diagram which has been created to convey a good understanding of FDM. In satellite communications our carrier is at microwave frequencies, which means that a small percent bandwidth of a high carrier frequency will result in a large frequency range in which many signals can be frequency-division-multiplexed; that is, if our carrier frequency is 8 GHz, a small 2 percent BW of this carrier frequency will equal 160 MHz. If, in our example, each channel requires 4 kHz, a total of 40,000 channels can be frequency-division-multiplexed onto a single carrier of 8 GHz.

Frequency Division Multiple Access (FDMA).   This uplink method involves a group of earth stations, each of which is assigned a particular frequency band. Each station transmits its information on its own particular frequency to the satellite. This multiplexed information is amplified, converted to the downlink frequency by the transponder, and then rebroadcasted back to the receiving earth stations. The receiving earth stations de-

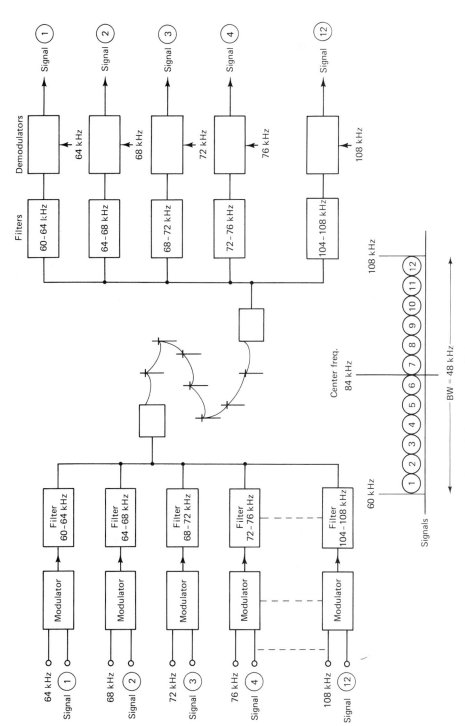

Figure 7-7 Frequency division multiplexing.

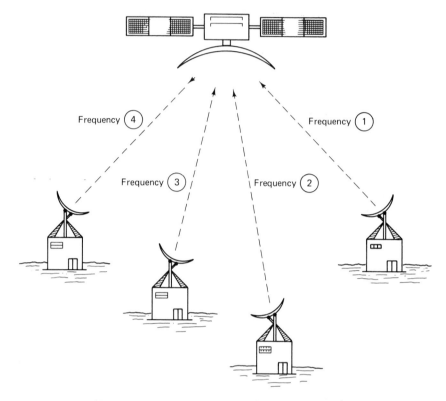

Figure 7-8   Frequency division multiple access (FDMA).

cipher this multiplexed transmission by tuning to whatever frequency their information was placed on. Figure 7-8 illustrates three earth stations using FDMA.

2. Time Division Multiplexing (TDM)

This process involves two or more terminals, each being allotted a position in time with reference to a synchronizing signal. Every terminal in a group transmits its signal information sequentially in different time periods to the receiving station; that is, time is split up and two or more signals from different terminals are multiplexed into segments of time.

Figure 7-9 shows a simple example of time division multiplexing 24 voice signals. The sampling gates are closed and opened one after another to allow the voice signal to pass through to the pulse modulator. Here the

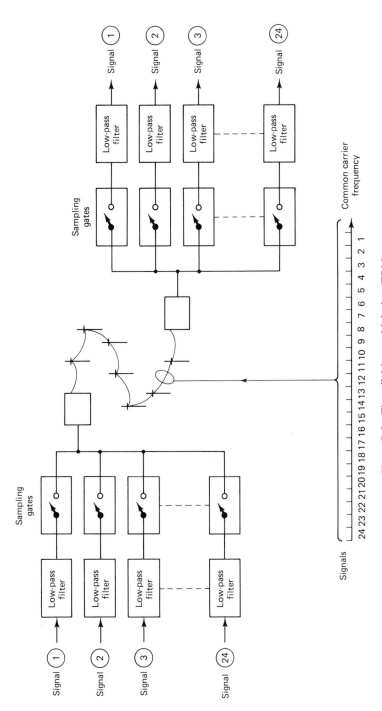

Figure 7-9 Time division multiplexing (TDM).

voice signal is encoded into pulse modulation, and then transmitted to the receiving station.

The pulse modulation decoder reconstructs the voice signal, and the sampling gates allow the correct voice signal to flow out of the correct sampling gate to the called party.

The sampling gates at the transmitter and receiver must obviously be in synchronization; that is, sampling gates 1, at the transmitter and receiver, are closed at the same time. Gates 2 are closed at the same time, and so on. This synchronization is achieved by sending synchronization information from transmitter to receiver, along with the voice signals.

To summarize, one signal after another is switched through, and every channel is separated from the next because it occupies a different time slot. Consequently, many signals are being transmitted on the same carrier frequency, but do not interfere with one another because they are separated in time.

Time division multiplexing is used to communicate:

a. Analog information (voice). This requires first encoding the voice signal into pulse modulation, and then transmitting this pulse modulation in an assigned time slot (TDM)

b. Digital information (teletypewriter). When a key is depressed on the keyboard, a digital code is generated, and these codes, from all the characters, will modulate a carrier during a time slot.

Let us now proceed to analyze these two points in a little more detail, starting first with pulse modulation of analog information.

PULSE MODULATION

Figure 7-10 illustrates a complex wave with a number of discrete plotted points. If only these points were given, the curve could be accurately drawn by guessing the path in between the points.

A pulse modulation encoder at the transmitter takes samples of a waveform at regular intervals and transmits these samples as pulses on a carrier.

A pulse modulation decoder at the receiver will reconstruct a replica of the transmitted wave by quite simply joining the dots.

Pulse modulation (PM) differs from amplitude and frequency modulation (AM and FM), in that the two latter methods continuously modulate the carrier, whereas in pulse modulation there is a time interval between

Figure 7-10   Complex wave with plotted points.

pulses. This time interval, instead of being wasted, can be occupied by other pulses from other signals that also need to be transmitted. This, therefore, allows a number of signals to be transmitted on the same carrier frequency at different times. Pulse modulation is therefore an ideal candidate for use in time division multiplexing.

Figure 7-11 shows a sine wave to be encoded and transmitted by using one of the three pulse modulation methods also illustrated.

1. Pulse Amplitude Modulation (PAM)

In this example the amplitude of the pulses are made proportional to the modulating signal's amplitude (the sine wave).

2. Pulse Duration Modulation (PDM)

Also known as pulse length and pulse width modulation (PLM and PWM). In this example, the amplitude of the pulses remains constant, but the width of the pulses vary dependent on the modulating signal.

3. Pulse Position Modulation (PPM)

This method again maintains the amplitude of the pulse constant, but the position of the modulated pulse (thick line) is varied dependent on each

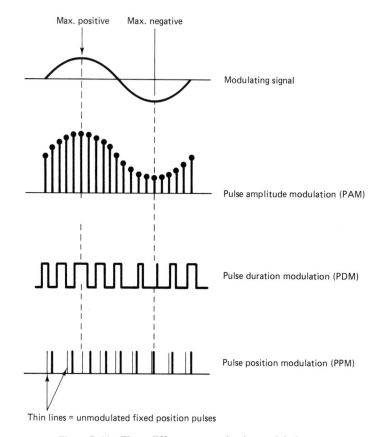

Thin lines = unmodulated fixed position pulses

Figure 7-11   Three different types of pulse modulation.

sample of the modulating signal. That is, the more positive the modulating waveform, the more the modulated pulse (thick line) moves away from the fixed-position reference pulses (thin line).

Figure 7-11 illustrates three examples of pulse modulation, but it greatly exaggerates the number of pulses required. It would obviously be ideal to have a very large number of pulses transmitted, as the more pulses transmitted, the better the replica at the receiver. However, the major application of pulse modulation is time division multiplexing, and fewer pulses transmitted means that more room is available for more multiplexed signals.

In fact, it can be mathematically proven that if the number of samples per second exceeds twice the highest frequency contained in the modulating signal, then the waveform will be reconstructed at the receiver with a good fidelity. Let us take this statement a stage further and prove this point with a typical voice waveform illustrated in Fig. 7-12.

The pulses transmitted, using PAM, will reconstruct the voice signal at the receiver with a reasonable degree of accuracy. The highest frequency component is found to be 4 kHz, so our sampling rate will be at a frequency of 8 kHz. This sampling rate is the minimum recommended rate for voice transmission, and since our PAM pulses will only occupy approximately 1 μs of time, a large amount of space exists between pulses where other voice transmissions could be multiplexed onto this same transmission. If the sampling rate is increased to improve fidelity, time between pulses is reduced, and fewer channels can be multiplexed on to the same transmission, as less room is available between consecutive pulses.

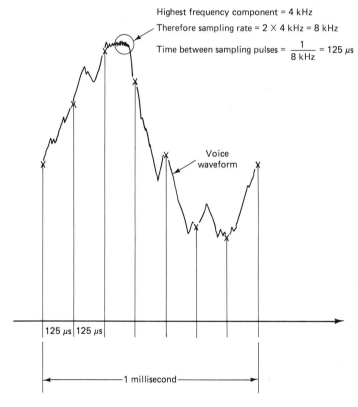

Figure 7-12 PAM of a typical voice signal.

MULTIPLE ACCESS

This is when a communication satellite is used by more than one pair of earth stations at a time; that is, many different earth stations use the satellite without interfering with one another by utilizing either FDM or TDM

FDMA (frequency division multiple access), which has already been discussed, and TDMA (time division multiple access) are the two prime methods of "uplink" transmission.

TDMA

This second method for satellite uplink is, as already discusssed, the multiplexing of many transmissions onto one carrier, but separating the transmissions in time, and therefore allowing only one package of information to reach the satellite at any instant in time.

Each station stores its information to be transmitted. A master station controls all the other earth stations, and tells them when to transmit their stored information and when to stop to allow the next station to transmit. The satellite actually receives a continuous signal made up of rapid bursts from many stations. This is illustrated in Fig. 7–13.

THE TELEPRINTER

The teleprinter or teletypewriter is a transceiver that can transmit and receive alphanumeric (letter of the alphabet and numerals) characters, in the form of separate codes for each one of these graphic symbols. When a key is depressed, a unique key code for that character is generated and transmitted as a train of pulses to the receiving station. The receiving teletype receives and decodes the code, and prints out the appropriate character. Both the transmitting and receiving teletypes produce a typewriter-like copy; the transmitter teletype produces its copy which it has just transmitted for checking and filing purposes. Figure 7–14 illustrates a teletypewriter.

If the operator has low typing speeds and efficiency, the message can be prepared in advance on a word processor or a perforated tape unit at the operator's speed. When correct, the message can be transmitted at maximum speed, providing maximum channel usage efficiency. The word processor allows the operator to enter in and continuously edit and correct the message until it is correct. The perforated tape unit allows the operator to

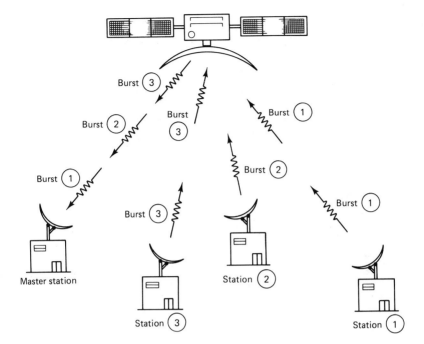

Figure 7–13

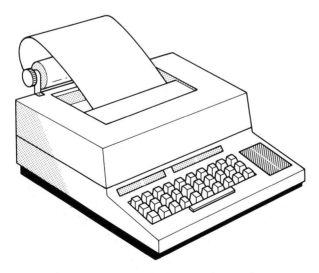

Figure 7–14   The teleprinter or teletypewriter.

type out the message at a convenient time and speed, producing codes on a perforated tape. The CCITT-2 code is illustrated in Fig. 7-15.

Every character in this code is made up of five binary digits (bits), in which a hole is referred to as a *mark* and the absence of a hole is a *space*. Each mark or space has a duration of 22 milliseconds (ms), so each and every 5-bit code occupies 110 milliseconds of time. Each 5-bit character code is preceded by a 22-ms space and followed by a 33-ms mark, and as the average word has 6 letters, the time needed to transmit an average word is equal to:

$$6 \times (22 \text{ ms} + 110 \text{ ms} + 33 \text{ ms}) = 990 \text{ ms}$$

Approximately one word per second is transmitted, which calculates out to be a speed of 60 words per minute.

A baud is the unit of signalling speed derived from the shortest code element duration. Speed in bauds is the number of code elements per second or frequency. The frequency or baud rate in our example can be easily calculated by taking the reciprocal of 22 ms, which calculates out to be a 45-baud system speed. This previous example explains the basic principle but is very slow, 1200 and 2400 baud systems are common place today.

Frequency Shift Keying (FSK)

As mentioned previously, a 5-bit code will be generated when a key is depressed. This code consists of a combination of marks and spaces that are unique to the particular character. For example, the 5-bit code for "A" is (refer to the CCITT-2 code in Fig. 7-15)

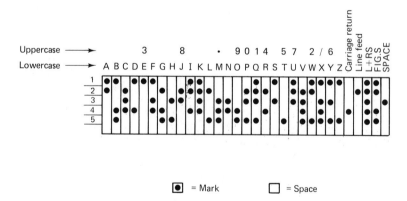

Figure 7-15   The CCITT-2 code.

$$\frac{1 \quad 2 \quad 3 \quad 4 \quad 5}{\text{mark} \ \text{mark} \ \text{space} \ \text{space} \ \text{space}} = A$$

Frequency shift keying is a form of frequency modulation (FM) in which the 5-bit code modulation wave shifts the output frequency between two predetermined values. A mark shifts the frequency above the carrier frequency; a space shifts the frequency below the center frequency, as illustrated in Fig. 7–16, which shows FSK of a 5-bit code for "P."

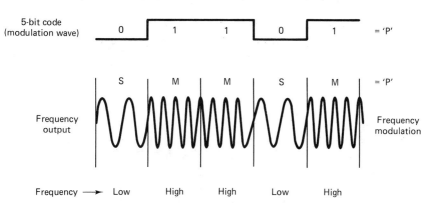

Figure 7-16   FSK for the letter "P."

Phase Shift Keying (PSK)

This relatively new system involves the modulating 5-bit code signal, shifting the carrier +90 degrees for a mark, and −90 degrees for a space. Figure 7–17 illustrates PSK of the 5-bit code for "P."

Both FSK and PSK achieve the same desired result, which is to take our 5-bit (mark/space) digital code from the teleprinter, and modulate a carrier so that it can be transmitted from *A* to *B* via satellite.

MULTIPLEXING TELEGRAPHY

FDM.   Telegraphy transmissions are normally sent over telephony channels, which have bandwidths between 300 to 3400 Hz (3100 Hz BW). So if, for example, 120 Hz is the bandwidth required for a telegraphy channel using FSK (60 Hz frequency shift and a 60-Hz guard band), 24 channels (24 × 120 = 2880 Hz) could easily be frequency multiplexed into the 3100 Hz bandwidth.

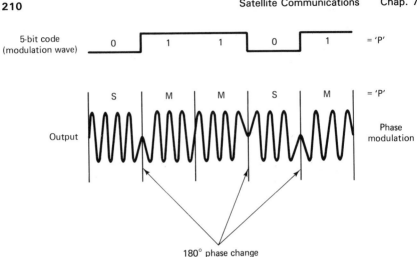

Figure 7-17   PSK for the letter "P."

TDM.   Time division multiplexing for multiple telegraphy transmissions is now very much established. Codes from various stations are allotted different positions in time, and as many as 128 different telegraphy transmissions could eventually be transmitted on one telephony channel using TDM.

THE SATELLITE SPACECRAFT

The satellite is a combination of rocketry and microwave electronics.

The microwave electronics is powered by solar cells and rechargeable nickel cadmium batteries which continue operation through the periods of eclipse with the earth. Continuous monitoring of the power available and the condition of the transponders is communicated to the earth station to ensure that the satellite is ready to respond to any earth station commands and has the power to carry out these commands. These condition signals also enable the earth station to track the satellite, and maintain all its fixed antennas directly on the satellite. The main failures within satellites are TWTs and power systems. Fuel often becomes a problem, as the satellites tend to naturally stray from their desired position, so the number of years these satellites remain active is dependent on how much fuel is expended to reposition the satellite. Consequently, designers have the difficult choice of deciding whether to incorporate more back-up electronics or fuel in the very limited space within the satellite.

ORBITS

Basically, celestial mechanics tells us that a satellite will orbit the earth at a velocity that is dependent on:

1. The satellite's distance from the earth. That is, a low-orbit satellite could orbit the earth in two hours, whereas the moon, which is in a far orbit of 385, 000 km away, orbits the earth in 28 days.
2. Whether the satellite is in a circular or low elliptical orbit.

Usually, satellites are placed in almost circular orbit just above the equator, with a 24-hour rotation speed.

A MARINE SATELLITE COMMUNICATION SYSTEM

There are three main components in a marine satellite communication system:

1. Geostationary Satellite

There are three of these satellites, one each over the Pacific, Atlantic, and Indian Oceans, positioned 22,300 miles above the equator. They serve as unattended relay transmitters for ship-to-shore and shore-to-ship communications. The coverage areas for these satellites are plotted on a map, which is illustrated in Fig. 5–14.

2. Shore Station

This earth station connects the land telephony and telegraphy (telex) networks to the satellite. It monitors a common request carrier frequency, awaiting request bursts from ships' terminals requesting the shore station to assign communication channels. An assignment message is part of the TDM (shore-to-ship) carrier, and is addressed specifically to the terminal that transmitted a request burst. When more than one shore station services the same ocean area, one of the earth stations will function as network coordinating station (NCS) for the traffic, and will route your call via the shoreside network that provides the shortest path.

3. Ship's Terminal

The terminal connects the teleprinter and telephones on board the ship to the earth station via the satellite. The shore station with an assignment message in a channel of its TDM carrier will remotely establish the communication when a call arrives that is addressed to the ship's terminal, or when requested to make a communication to shore.

FREQUENCIES

Duplex communication (two-way, one in each direction) can be achieved for either telephony or telegraphy by using two carrier frequencies in both directions. In the MARISAT/INMARSAT system the frequency pairs are arranged with the receiving (shore-to-ship) downlink frequency 101.5 MHz below the transmitting (ship-to-shore) uplink frequency. Figure 7–18 shows the uplink and downlink frequenices for:

1. Shore-to-ship. Telegraphy (TDM) and telephony (voice), downlink.
2. Ship-to-shore. Telegraphy (TDMA) and telephony (voice), uplink.

MARISAT/INMARSAT reserved 339 carrier frequencies, spaced at 25 kHz increments. Out of these frequencies our system will use:

1. 8 for shore-to-ship telegraphy for TDM.
2. 8 for ship-to-shore telegraphy for TDMA.

TDM and TDMA frequencies are always used in pairs.

The MARISAT satellites above the Atlantic and Pacific had the capacity for eight voice channels, whereas the Indian Ocean satellite had only two voice channels. The new INMARSAT satellites, which came into operation in 1982, have the capacity to handle approximately 45 equivalent telephony channels.

SHIP-TO-SHORE COMMUNICATION

a. Telegraphy

Step 1 (Fig. 7–19).   The operator on board the ship types three digits on the teleprinter, and in so doing activates the ship's terminal, which sends a request burst specifying the type of communication desired (te-

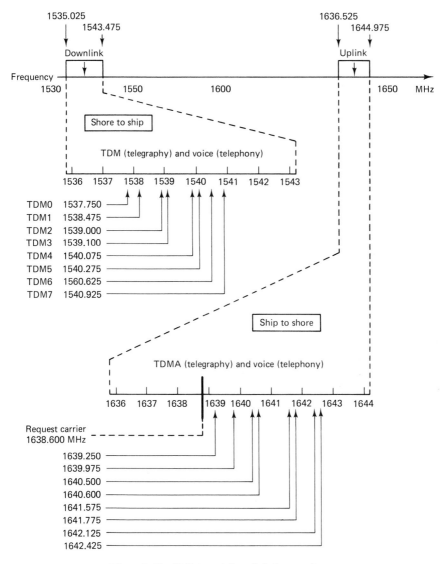

Figure 7-18 Uplink and downlink frequencies.

lephony or telegraphy). The request burst is transmitted on a common carrier frequency of 1638.6 MHz, which is common to all terminals. Only one request burst (frame) is necessary, and the access to the shore station is on a first-come, first-served basis. Figure 7–20 illustrates the signal format for the request from ship to shore.

All relevant specifications of the call must be entered in the infor-

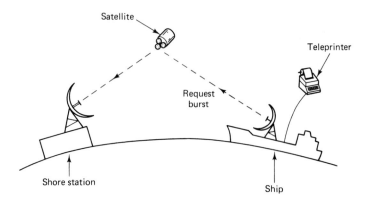

Figure 7-19

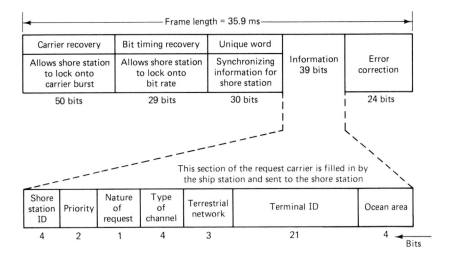

Figure 7-20   Signal format for ship-to-shore "request burst."

mation field of the request burst. This type of signalling before a connection has been established is referred to as *out-of-band signalling*.

Step 2 (Fig. 7–21).   The shore station sends an assignment message to the ship terminal via the assignment channel of the TDM carrier. This assignment message specifies the shore-to-ship TDM frequency for reception and the ship-to-shore TDMA frequency for transmission, and then tunes the ship's terminal to these frequencies. Furthermore, this assignment message also assigns a particular time slot in the TDM and TDMA formats, and sets up a demultiplexer (for reception of TDM), and multiplexer (for transmission of TDMA) within the terminal for the particular time slot, so our terminal will only transmit and receive during our assigned time period. In the time division multiplexing employed for the TDM and TDMA carriers, the 22 telegraphy channels share the same carrier by using it one after another in rapid succession. Let us take a closer look at the signal formats for the TDM and TDMA. Figure 7–22 illustrates the shore-to-ship TDM format.

TDM.   One complete scan is referred to as one frame, which for the TDM carrier lasts 0.29 second.

To gain a time slot, the ship's terminal will have already released a request burst, and the shore station will send an assignment message addressed to the terminal in the assignment channel of the TDM carrier. Only if the ship's terminals own identification is recognized in the terminal ID slot of the assignment channel will it pay attention to the rest of the assignment message where the assigned time slot is specified.

The time interval for one channel in each frame is denoted as a time slot, and has a fixed time relation to the synchronizing word (unique word), which is transmitted at the beginning of each frame. This unique word is

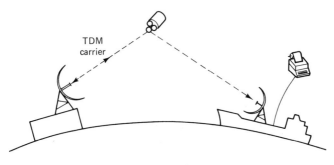

TDM
carrier

Figure 7-21

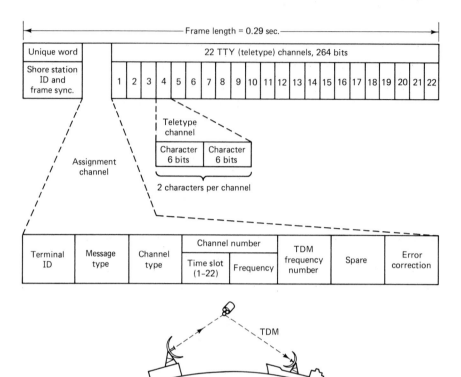

Figure 7-22 Shore-to-ship TDM format.

transmitted at the beginning of each frame. This unique word is used by all the ships' terminals (that have a time slot) as a time reference to start a counter within the ship's terminal. At the count corresponding to the assigned time slot in the TDM, the ship's terminal opens up and lets the received signals through for only that time period.

In our example, illustrated in Fig. 7-22, we are receiving two 6-bit codes during the fourth teletype channel on TDM.

TDMA. The TDMA (ship-to-shore) format is illustrated in Fig. 7-23.

The time division multiple access, TDMA, in the opposite direction from ship-to-shore, also contains 22 telegraphy channels, but the format differs from that of the TDM multiplex, and the frame is six times longer.

The unique word of the TDM carrier is inverted every sixth TDM frame, which will not have any affect on the system's operation, but will be used for timing of the TDMA carrier. As the TDM frame length is equal to 0.29 second, the inverted unique word will occur after six TDM frames;

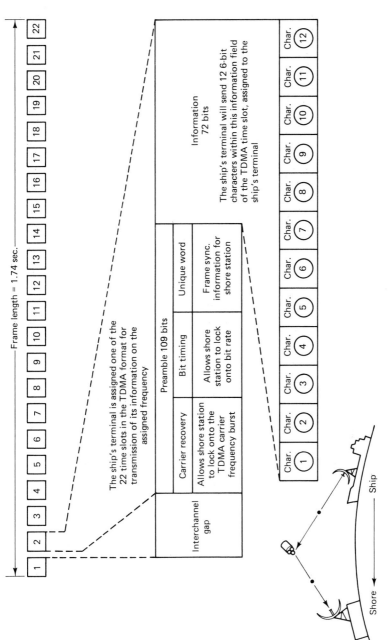

Figure 7-23 Ship-to-shore TDMA format.

that is, 6 × 0.29 second = 1.74 seconds (the TDMA frame length). The beginning of each TDMA frame for all terminals is locked to the inverted unique word of the received TDM signal.

In the example, illustrated in Fig. 7–23, we have been assigned the second time slot to transmit our twelve 6-bit codes.

The interchannel gaps between the time slots of the TDMA format permit a certain timing tolerance (overlap into another channel or time slot) caused by variation in the propagation for ships at various distances from the shore via satellite.

Step 3 (Fig. 7–24).   While the communication is in progress on the assigned channel, the ship's terminal continues to listen to the TDM carrier for other possible messages addressed to the terminal.

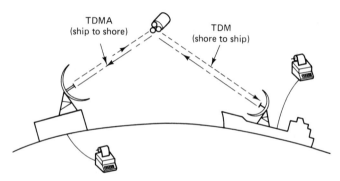

TDMA
(ship to shore)

TDM
(shore to ship)

Figure 7–24

b. Telephony

Step 1 (Fig. 7–25).   The initialization is almost identical to telegraphy, in that the ship's terminal sends a request burst by dialing three

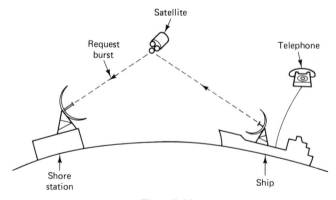

Satellite

Request
burst

Telephone

Shore
station

Ship

Figure 7–25

digits. The request burst specifies the type of communication to follow (telephony), and the ship's terminal then waits for the shore station to assign a voice carrier pair for a duplex (ship-to-shore) communication.

Step 2 (Fig. 7–26).   The shore station sends the assignment message addressed to the ship's terminal in the assignment channel of the TDM carrier. This assignment message will specify the voice carrier pair to be used and will tune the terminal to these frequencies.

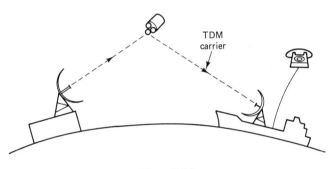

Figure 7–26

Step 3 (Fig. 7–27).   Each voice carrier is frequency modulated (FM) by one voice signal only with a bandwidth of 3 kHz, which is adequate for telephony.

While the communications is in progress on the voice carrier pair, the terminal continues to listen to the TDM carrier for other possible messages addressed to the terminal.

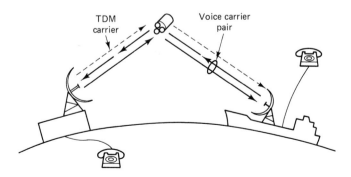

Figure 7–27

BLOCK DIAGRAM OF A SHIP'S SATELLITE
COMMUNICATION TERMINAL

Figure 7–28 shows a block diagram of a typical ship's terminal.

The antenna is a 4-ft paraboloid with a gain of approximately 23.5 dbs, and is mounted on a stabilized platform which automatically compensates for pitch and roll movement of the ship.

The antenna drive unit has to adjust the antenna in bearing (azimuth) and elevation as the position and heading of the ship change.

A tracking control system monitors the received signal strength of the TDM carrier from the shore station, and uses it as a reference for automatic tracking of the satellite. That is, the tracking control block monitors the antenna's position signals (azimuth and elevation), the gyro (from the ship's compass when the ship changes course), and the TDM signal strength. It then feeds all this information to the function controller, which samples all these signals to gain an overall picture and then generates the required drive commands for the drive motors in the antenna unit via tracking control block to maintain the antenna locked on to the satellite.

The antenna can alternatively be positioned by manual controls on the front panel below deck.

The RF unit above deck is split into a transmitter and receiver section:

1. Receiver (downlink) 1535 to 1543.4 MHz

The received satellite signals (one of the eight TDM telegraphy carriers or a shore-to-ship voice carrier) are fed via the LNA (low noise amplifier) to the first down converter, which transposes the 1500-MHz range signals to a VHF frequency range of 200 MHz. The injection frequency (1330 MHz) for the down conversion is generated from an injection frequency oscillator (IFO), which receives a 5-MHz reference frequency from the central terminal unit (CTU) below deck. The frequency conversion in the down converter and the injection frequency oscillator occur in the RF unit above deck rather than below, so lower losses occur in the coaxial cables between antenna and CTU. In other words, you can place the CTU in the most convenient location without worrying about high losses in the microwave signal path, because our signals are no longer on microwave frequency carriers and therefore will travel in coaxial cables with very little loss.

Our received signals on a VHF carrier frequency are coupled down to the CTU, which contains the remaining RF circuits and all the digital circuits that control the operation of the terminal.

Tuning to the desired received signal is accomplished in a second down converter, where one of many injection frequencies from a synthesizer bring

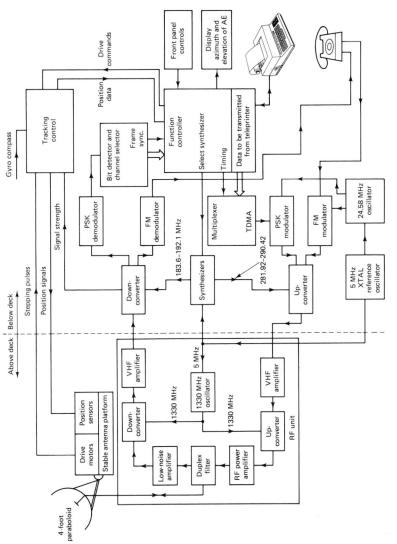

Figure 7-28 Marine satellite communications terminal.

the desired signal down onto an intermediate frequency (IF) of 21.4 MHz. If, for example, the signal we desired was on a carrier frequency of 205 MHz as it entered the down converter, the function controller would select a synthesizer frequency of 183.6 MHz, which when mixed with 205 MHz in the down converter would produce a sum and difference. The difference (IF) frequency of 21.4 MHz would be allowed to pass through the FM demodulator (if telephony) or PSK demodulator (if telegraphy). If, on the other hand, we wish to tune in a signal which sits on a VHF carrier of 213.5 MHz as it enters the down converter, the function controller will select a synthesizer frequency of 192.1 MHz which again will allow only through the down converter, the signal that entered on a frequency 21.4 MHz above the injection frequency (from the synthesizer) of 192.1 MHz, that is, 213.5 MHz.

There are separate synthesizers and separate down converters for the:

a. TDM carriers, which are applied to the PSK demodulator, which extracts the TDM format from the PSK IF.

b. Voice carriers, which are applied to an FM demodulator, which removes the voice signal from the IF and sends the voice signal to the telephone, where it is heard in the earpiece.

In the idle condition the CTU is tuned for reception of the TDM carrier, which is subsequently applied to the PSK demodulator, where the TDM signal is recovered. This received TDM signal is fed to the bit detector and channel selector block, which looks for the frame sync (unique word) within the TDM format. When the frame sync is located, the channel selection block informs the function controller. Once the function controller is aware of when the frame sync occurs, it will be able to extract the terminal's channel information (two 6-bit teletype characters) from the appropriate time slot and feed it to the teleprinter.

The assignment message within the TDM format is always monitored by the function controller, which decodes the message and takes the required actions, that is, selects a time slot and synthesizer frequency for TDM reception and TDMA transmission. Once a TDMA frequency and time slot have been assigned, the function controller turns on and off the transmitter at the appropriate time.

2. Transmitter (uplink) - 1636.5 to 1645.0 MHz

a. Telephony. The voice carrier pair has already been assigned, and the function controller has selected the synthesizer frequencies that will be injected into the up and down converters to achieve reception and transmission on the assigned voice-carrier pair. The voice signal to be transmitted is applied to a frequency modulator from the telephone's microphone, and this signal frequency modulates an intermediate frequency (IF) of 24.58 MHz. The injection frequency range from the synthesizers is from 281.92 to 290.42 MHz, and when it is mixed in the up converter with the 24.58 MHz IF, a resulting VHF frequency range exists between 306.5 to 315 MHz which carries our voice signal from below to above deck.

b. Telegraphy (TDMA). The function controller sends the data from the teleprinter to the multiplexer block, where it is held under the control of the function controller. When the assigned TDMA time slot occurs, a timing signal from the function controller tells the multiplexer to send the twelve 6-bit telegraphy characters to the PSK modulator. A 24.58-MHz reference and the TDMA signal from the multiplexer are applied to the PSK modulator. A PSK-modulator carrier from the PSK modulator block is then up-converted by means of the injection frequency from the synthesizer to within the VHF carrier range of 306.5 to 315 MHz.

In the RF unit above deck, our signal (either telephony or telegraphy) is amplified and then up-converted by mixing our VHF frequency from below deck with an injection frequency of 1330 MHz, which produces our final microwave uplink frequency. In telegraphy this will be one of the eight TDMA frequenices and in telephony a ship-to-shore voice carrier frequency.

The RF power amplifier boosts our voice signal and is turned on and off in the telegraphy to ensure that we transmit only on the TDMA carrier in our assigned time slot.

The final signal at the microwave uplink frequency is fed via the duplex filter, which is used to isolate the transmitter and receiver to the antenna and then to the shore station via satellite.

SUMMARY

1. Generally a frequency difference of 2 GHz exists between uplink and downlink to prevent interference between the two.
2. The INTELSAT satellites provide point-to-point transoceanic communications.
3. The INMARSAT satellites provide communications from ship-to-shore and vice versa.

4. Frequency division multiplexing is when the available bandwidth is divided into narrow bands, each of which is used to transmit a separate channel.

5. Time division multiplexing is when time is split up and two or more signals from different sources are transmitted at different times on the same carrier.

6. Pulse modulation is used in time division multiplexing.

7. The teleprinter or teletypewriter is a transceiver that can transmit and receive alphanumeric characters.

8. Frequency and phase shift keying are modulation methods used to transmit five-bit codes from the teleprinter.

REVIEW QUESTIONS

1. The INTELSAT satellite organization provides:
   (a) Communication between ship and shore and vice versa.
   (b) U.S. regional communication
   (c) Transoceanic communication
   (d) None of the above

2. The INMARSAT satellite organization provides:
   (a) Communication from ship to shore and vice versa
   (b) U.S. regional communication
   (c) Transoceanic communication
   (d) None of the above

3. Name three U.S. regional satellites
   (a) _____
   (b) _____
   (c) _____

4. Briefly describe what is meant by a multiplexed communication.

5. Give the full names for the following abbreviations:
   (a) TDM
   (b) FDM
   (c) FDMA
   (d) PAM
   (e) PDM
   (f) PPM

6. Briefly describe what is meant by *multiple access*.

7. What are the two prime uplink methods?
   (a) ——————————————
   (b) ——————————————

8. What is the five-bit CCITT-2 code for each of the following?
   (a) Z
   (b) H
   (c) line feed
   (d) space
   (e) D

9. What are the three main components in a marine satellite communication system?
   (a) ——————————————
   (b) ——————————————
   (c) ——————————————

10. The request burst is sent from ship-to-shore on a carrier frequency of ——————————————. Refer to Fig. 7–28 for the following six questions.

11. The antenna is mounted on a stabilized platform which:
    (a) Determines the antenna's gain
    (b) Has no effect on anything
    (c) Compensates for ship's pitch and roll
    (d) None of the above

12. The tracking control system provides automatic tracking by:
    (a) Monitoring the strength of the TDMA carrier
    (b) Monitoring the strength of the TDM carrier
    (c) Use of the manual tracking controls
    (d) None of the above

13. What type of shift keying is used for the telegraph signals:
    (a) FSK
    (b) PSK
    (c) FM
    (d) PPM

14. What is the function of the duplex filter?

15. Which block is the brain of the system?
    (a) The up-converter
    (b) The tracking control
    (c) The function controller
    (d) The synthesizers

16. All frequencies for the whole system are derived from which fundamental frequency block:
    (a) 1330-MHz oscillator
    (b) 5-MHz oscillator
    (c) 24.58-MHz oscillator
    (d) All of the above

17. A voice carrier is _____ and each voice signal has a bandwidth of _____ .
   - **(a)** Frequency modulated, 3 kHz
   - **(b)** Amplitude modulated, 3 kHz
   - **(c)** Frequency modulated, 5 kHz
   - **(d)** Amplitude modulated, 30 kHz

18. How many characters can be sent in the information field of the TDMA carrier?
   - **(a)** 14
   - **(b)** 6
   - **(c)** 2
   - **(d)** 12

19. What is the TDMA frame length?
   - **(a)** 0.29 sec
   - **(b)** 1.74 sec
   - **(c)** 35.9 ms
   - **(d)** 13.47 sec

20. Every sixth TDM unique word will be inverted and used for timing the start of the:
   - **(a)** TDM carrier
   - **(b)** TDMA carrier
   - **(c)** request burst
   - **(d)** system shutdown

GLOSSARY

Analog data A representation of information which bears an exact relationship to the original information. Electrical signals in a telephone channel are an analog data representation of the voice signal.

Automatic frequency control A system that produces an error voltage proportional to a frequency drift. The error voltage causes an oscillated change in the reverse direction to correct the original drift.

Azimuth The angular measurement in the horizontal plane bearing.

Beam width The angular width of a radar beam measured between the lines of half-power intensity.

Bearing resolution The minimum angular separation in the horizontal plane between two targets that will produce separate echoes (directly proportional to horizontal beamwidth).

Binary A base 2 number system in which two digits (1 and 0) are used.

Cavity resonator A space in which oscillating electromagnetic energy occurs within an electrically conducting surface. The resonant frequency is determined by the physical size of the enclosure.

Digital communications The transmission of information by use of encoded numbers, that is, binary.

Digital data Data usually represented by means of coded characters.

Duplex A communication in which either end can simultaneously transmit and receive.

Duty cycle The amount of time that a device is operating, as opposed to its idle time.

Frequency division multiplexing Taking the frequency spectrum or bandwidth for one transmission and subdividing it into separate channels, each of which will transmit data from a separate source.

Frequency shift keying A form of frequency modulation in which the modulating wave shifts the output frequency between two points when sending binary coded characters.

Horn antenna A rectangular antenna that is wider at the open end.

Klystron An electron tube used as an oscillator or amplifier at microwave frequencies.

Magnetron An electric tube used to generate high-power microwave frequency outputs.

Multiple access The use of a communication satellite by more than one pair of ground stations at a time.

Multiplex operation Simultaneous transmission of two or more messages.

Paraboloid reflector A hollow concave reflector, also known as a *dish*.

Phase shift keying A form of phase modulation in which the modulating wave shifts the instantaneous phase of the output between two predetermined points.

Plan position indicator A radar presentation in which the surrounding situation appears on the screen in plan form.

Precipitation attenuation A reduction in energy as it passes through the atmosphere that contains precipitation. The loss can be either scattering or absorption.

Propagation The travel of electromagnetic waves through a medium

Pulse amplitude modulation The modulating wave amplitude modulates a pulse carrier.

Pulse duration modulation The modulating wave varies the width of a pulse.

Pulse modulation A series of modulated pulses that carry and convey information.

Pulse repetition frequency The rate of the pulses.

Pulse repetition time The time between pulses.

Pulse position modulation The modulating wave is used to vary the position in time of a pulse.

Range resolution The smallest difference in range between two targets on the same bearing that will produce separate echoes.

Reflex klystron A klystron with a repeller and resonant cavity that is used as an oscillator.

Subrefraction Atmospheric refraction that is less than standard.

Super refraction Abnormal refraction causing a range increase.

Synthesizer A device that can generate many crystal control frequencies.

Telegraphy A system of telecommunication for the transmission of alphanumerics and signal codes.

Telephony Speech transmission.

Teletypewriter Also referred to as *teleprinter*. A keyboard machine that can transmit and receive alphanumeric and signal codes.

Thermionic emission The emission of electrons by temperature.

Time division multiplexing Two or more channels of information transmitted on the same carrier at different time intervals, that is, interleaved.

Transponder A receiver/transmitter which receives and then retransmits information.

Travelling wave tube amplifier   A power amplifier used to amplify microwave signals.

T/R switch or tube   A gas-filled switching tube that allows a common antenna for transmission and reception.

Waveguide   A hollow conducting device used to guide high-frequency electromagnetic waves.

Waveguide cutoff frequency   The lower frequency limit of propagation down a waveguide.

ANSWERS TO ODD-NUMBERED QUESTIONS

Chapter 1

1. (a) 5cm
   (b) 1.7cm
   (c) 2.4mm
   (d) 1.1mm

3. (a) 3.3 nanoseconds
   (b) 3.3 picoseconds
   (c) 16.6 picoseconds
   (d) 4.16 picoseconds

5. (b) Centimeters

7. (a) 186,000 miles/second
   (b) 300,000,000 meters/second

9. (a) Extra high frequency
   (b) Super high frequency
   (c) Ultra high frequency
   (d) Identify friend or foe
   (e) Hertz
   (f) Very high frequency

11. (b) 2.45 GHz

13. (a) amplitude
    (b) velocity
    (c) frequency
    (d) wavelength

15.   300 MHz to 300 GHz

Chapter 2

1. (a) direct wave
   (b) reflected wave
   (c) ground wave
   (d) sky wave
3.   direct and reflected
5. (c) faster, downwards
7. (a) The direct wave
9. (c) Scattering
11.   electric, magnetic
13. (d) All of the above
15. (c) sub-refraction
17.   10 meters
19. (a) temperature
    (b) precipitation
    (c) atmospheric pressure

Chapter 3

1. (d) TE_{10}
3. (e) Waveguide
5. (a) A resonant cavity
7. (d) Both (a) and (c)
9. (a) Detector or mixer
11. (d) All of the above
13. (a) Pyramid
    (b) Vane
15.   True
17.   high
19. (a) Buncher
    (b) Catcher
21. (a) Only one cavity
    (b) Internal feedback
23. (a) The length of the tube
25. (a) Thermistors
    (b) Barreters
27. (b) A negative temperature coefficient
29. (a) A positive temperature coefficient

Chapter 4

1. 5.37 db's
3. **(c)** 100
5. **(a)** Rectangular
   (b) Pyramidal
   (c) Conical
7. **(d)** All of the above
9. **(a)** Has slots cut in the waveguide to disturb current
11. **(b)** The horizontal beamwidth
13. **(a)** Rear-fed
15. **(d)** Large, small
17. **(a)** Reduces weight and **(b)** wind resistance
19. **(c)** Marine radars
21. **(b)** A radiating slot
23. **(a)** Nonradiating slot
25. slotted waveguide

Chapter 5

1. **(a)** 1100 feet per second
3. **(a)** Transmitter
   (b) Receiver
   (c) Display unit
   (d) Antenna
5. **(a)** 312.5uS
   (b) 3.125uS
   (c) 1200uS
7. Magnetron
9. **(b)** Correct bearing of target is displayed on PPI
11. **(d)** All of the above
13. **(d)** All of the above
15. **(a)** Reliability
    (b) Access time
    (c) Distress ease
    (d) Simple operation
    (e) Privacy

Chapter 6

Introduction Section

1. (b) Is the master timing device of the radar
3. (a) Timebase
   (b) Bright-up
   (c) Range rings
   (d) Variable range marker
5. (c) SWG, 3
7. (b) Difference between
9. (c) 8 nm
11. (a) scanner unit
   (b) display unit
   (c) rectifier unit
13. (c) Scanner and transceiver
15. Plan position indicator

Section One

1. (c) 440
3. (a) TX = synchro transmitter
   (b) TR = synchro receiver
   (c) Tx = transmitter
   (d) Rx = receiver
5. (c) Deflection coil, display unit
7. (b) 1
9. (d) Bearing of targets displayed

Section Two

1. (a) Pulse length
   (b) Pulse duration
   (c) Pulse repetition frequency
   (d) Pulse repetition time
   (e) Average power
   (f) Pulse width
   (g) Peak power
3. (d) All of the above

5. (a) True
7. (e) Both (a) and (c)
9. (b) Vary the radar's PRT
11. (c) 1500Hz, 0.08μs
13. (b) Inhibiting the charge trigger
15. (a) 666μs

Section Three

1. (d) 60MHz
3. (a) 60MHz above the echo frequency
5. (b) At short ranges
7. Automatic frequency control
9. (b) Local oscillators

Section Four

1. (a) Trigger
3. (b) Causes the beam to sweep from the center to the edge of the screen
5. Electronic bearing marker
7. It is used to show whether the radar is working at full efficiency or not.
9. (b) Scanner and deflection coil misalignment
11. (d) A/C rain control
13. (e) Both (a) and (c)
15. (a) Echoes
     (b) V.R.M. pulse
     (c) Tuning indicator pulse
     (d) Range rings
     (e) Heading marker

Chapter 7

1. (c) Transoceanic communication
3. (c) SATCOM
     (b) GSTAR
     (c) WESTAR
5. (a) TDM = Time division multiplexing
     (b) FDM = Frequency division multiplexing

    (c) FDMA = Frequency division multiple access
    (d) PAM = Pulse amplitude modulation
    (e) PDM = Pulse duration modulation
    (f) PPM = Pulse position modulation

7. **(a)** TDMA
   (b) FDMA

9. **(a)** Geostationary satellite
   (b) Shore station
   (c) Ships terminal

11. **(c)** Compensates for ship's pitch and roll

13. **(b)** PSK

15. **(c)** the function controller

17. **(a)** Frequency modulated, 3kHz

19. **(b)** 1.74 Sec

INDEX